AF505650

POINT PROCESS MODELS OF
CAVITY RADIATION AND DETECTION

Point Process Models of Cavity Radiation and Detection

A Statistical Treatment of
Photon Population Point Processes

S K SRINIVASAN

Professor of Applied Mathematics
Department of Mathematics
Indian Institute of Technology, Madras

CHARLES GRIFFIN & COMPANY LTD
London

OXFORD UNIVERSITY PRESS
New York

CHARLES GRIFFIN & COMPANY LIMITED
16 Pembridge Road, London W11 3HL, Great Britain

First published 1988

Published in USA by
Oxford University Press
200 Madison Avenue, New York, NY 10016, USA

OUP ISBN 0-19-520636-3

British Library Cataloguing in Publication Data

Srinivasan, S.K.
   Point process models of cavity radiation
   and detection: a statistical treatment
   of photon population point processes.
   1. Cavity resonators—Mathematical
   models   2. Stochastic processes
   I. Title
   621.381′332   TK7876

ISBN 0 85264 291 1

Typeset in Great Britain by
Kingprint International Limited, Richmond, Surrey
Printed & bound in Great Britain by
Redwood Burn Limited, Trowbridge, Wilts

Preface

The object of this monograph is to provide a general exposition on Markov and non-Markov modelling of cavity radiation by the combined application of point process theory and the techniques of branching. Most of the features of the models and the results deduced therefrom have not appeared in book form; more particularly the non-Markov evolution of the cavity photon population and the detection process have not been discussed in the literature. The material is addressed mainly to applied probabilists (who wish to take to the modelling of modern optical phenomena and lasing action) and optical scientists who wish to learn more about the potential use of stochastics. Particle physicists engaged in interpreting the results of colliding beam experiments may also find the book useful, especially in view of the revived interest in shower theory. The bulk of the material is an improved version of the notes of lectures delivered in the first half of 1986, mainly to students in the broad area of Applied Probability and Stochastics; however, among the audience there were professional colleagues and friends in constant interaction with the blackboard demonstrations and I owe them the idea that these lectures could be converted into book form.

Cavity radiation and detection has a long history. In 1917, Einstein studied the problem of black body radiation from general considerations of thermal equilibrium. His treatment of stimulated and spontaneous emission of radiation by molecules still dominates the thinking of physicists. However, it was much later that the idea that radiation could be amplified by cavity confinement gained currency, leading to the construction of the first maser device in 1954 by Gordon, Zeiger and Townes. About the same time, Hanbury-Brown and Twiss measured the intensity correlation of light from stars. These two developments provided a good deal of motivation for the study of light amplification and detection. A theoretical model dealing with the evolution of photon population was soon put forward by Shimoda, Takahasi and Townes[*] in 1957 and this accounted for many of the interesting properties of amplifiers. This stimulated further attempts, particularly by Mandel and Wolf, Scully and

Lamb, and Haken, culminating in a fully quantum-mechanical theory of light amplification. Simultaneously the problems of coherence and quantum detection received attention, particularly by Glauber and Sudarshan. However, it is pertinent to observe that although the theory of photoelectric effect due to Einstein provided a great stimulus to the development of quantum theory, light itself was studied until about 20 years ago only by recourse to classical and at best semi-classical methods, and it is only in recent times that a fully quantum mechanical theory of light is pursued seriously. More recently, the focus is on the non-classical behaviour of light and in particular the antibunched nature of the statistics of photo counts. Many interesting questions arise when the equation of evolution is analysed with reference to the photon population and its detection; there is an additional element of coarse graining and time-scales when problems of amplification are fully analysed, and these in turn introduce further statistical elements into the theory. In this monograph we confine our attention to some of the problems that arise when the population of photons is in interaction with the cavity atoms and the detector, and in so doing we mainly follow Shimoda, Takahasi and Townes. In addition we integrate some recent findings on the basis of non-Markov evolution. Such a treatment might appear very comprehensive and promising; however, it is severely constrained by practical considerations like the optimal size of the monograph and the author's personal inadequacies. Nevertheless it is hoped that the stochastics of the processes involved and the interesting conclusions that stem therefrom will be of interest to both students and researchers in the broad area of Applied Probability and Optical Science and Theoretical Physics in general.

The material is so arranged that the first four chapters can form the core of a short course on special topics of probability for students of physics and Electrical engineering. Throughout, the emphasis is on the use of backward differential equations adapted for regenerative processes of the branching type. The results presented in later chapters are of direct interest to optical scientists and engineers and are expected to stimulate further interest in theoretical research in estimating noise in optical devices. Biomathematicians and biostatisticians may also find the bulk of the material interesting, especially the chapters illustrating evolution through phases. Those who have no background of quantum mechanics/radiation theory may still find the bulk of the book consisting of Chapters 2, 3, 4, 6, 7 and 8 comprehensible; the introductory and concluding chapters may give them a general idea as to how stochastic population processes can be used to model optical phenomena.

The author owes to his teacher Professor Alladi Ramakrishnan and the late Dr S. Pancharatnam suggestions made in the 1960's for examining partially coherent light, and more particularly its detection on the basis of stochastic theory. The author's sustained research interests in the broad area of light and its detection are naturally due to the stimulus provided by his students and in particular the late Dr S. Sukavanam. The author also acknowledges with pleasure the active collaboration with his professional colleague Professor R. Vasudevan over decades, albeit punctuated by breaks. In particular the work on non-Markov modelling in recent times would not have been possible but for his active collaboration. The author is very grateful to his colleague Professor R. Subramanian for reading through the entire manuscript and making many useful suggestions.

In the past, preparation of the typescript had always been full of problems and the author had to go through a traumatic experience. This time he had the good fortune to feel relaxed all through, thanks to the arrangements for typing made by Professor R. Subramanian in his capacity as Head of the Department. The author is thankful to K. C. Venkatanarayanan of MSRC for the line drawings made at short notice. Finally thinks are due to V. Sridharan for his painstaking efforts in preparing notes of the lectures and his further help in proof-reading of the final typescript and preparation of the index.

Madras 1988   S. K. SRINIVASAN

(*) The population process model is exactly on the lines envisaged by J. Hans D. Jensen long ago. Charles Townes (*IEEE Journal of Quantum Electronics* **QE-20** 547–50: Ideas and Stumbling Blocks of Quantum Electronics) recalls a conversation with Jensen who told him that in the 1930s he thought of stimulated emission from an inverted population as a cascade of independent photons as in a cosmic ray shower, and that he had lost interest in the idea after an experiment which seemed to produce such effects turned out to be explainable otherwise.

Contents

1. INTRODUCTION   *page*   1–3

2. ELEMENTS OF STOCHASTIC PROCESSES   4–22
    2.1 Stochastic processes   4
    2.2 Markov processes   11

3. POINT PROCESSES AND PRODUCT DENSITIES   23–51
    3.1 Population point processes   23
    3.2 Moment measures: product densities   26
    3.3 Stationary point processes   35
    3.4 Cox processes   42

4. POPULATION GROWTH PROCESS   52–69
    4.1a Population model with constant rates: formulation   52
    4.1b Differential equation governing the generating functions and   54
           their solutions
    4.1c Steady-state characteristics   55
    4.1d Point process of emigrations: product densities   56
    4.2   Age-dependent population growth: Bellman–Harris process   59
    4.3   Age-dependent Kendall process: method of phases   61
    4.4   Kendall process: moments and age structure   65

5. CAVITY RADIATION   70–88
    5.1 Quantum approach to cavity radiation: density matrix   71
    5.2 Quantum evolution equations: coarse graining and Markov property   76
    5.3 Detection process   79
    5.4 Coherence functions and quantum detection process   83

6. A MARKOV MODEL OF CAVITY RADIATION   89–111
    6.1 Cavity evolution and equilibrium distribution   89
    6.2 Detection process: approach to equilibrium   93
    6.3 Detection process: point process of photoemissions   99
    6.4 Dead-time corrections to photoemission process   105

7. NON-MARKOV CAVITY RADIATION—   112–143
1: BELLMAN–HARRIS PROCESS

7.1   General characteristics of evolution   112

7.2   Detection process   116

7.3   Special models   118

7.4   Inhibited immigration   123

7.5   Two-stage model of evolution and emission: cavity population characteristics   127

7.5a   Analysis of the model   129

7.5b   Moment structure   132

7.5c   Behaviour of the second moment: Special cases   136

7.6   Two stage model: detection process   138

8. NON-MARKOV CAVITY RADIATION II: KENDALL PROCESS   144–167

8.1   Modified Kendall evolution: phase approach   144

8.1a   Evolution in phases   146

8.1b   Moments of the cavity population   147

8.1c   Behaviour of the third moment   150

8.1d   Arbitrary number of phases   151

8.2   Modified Kendall evolution: detection process   152

8.3   General phase-model: evolution through four phases   158

8.3a   Generating functions   158

8.3b   Factorial moments and cross-correlations   159

8.3c   A special model: superposition of two fields   162

8.4   Anti-bunching and sub-Poissonian characteristics   164

9. A BRIEF SUMMARY AND OUTLOOK   168–171

REFERENCES   172–176

INDEX   177–179

1. Introduction

Stochastic modelling of cavity radiation and detection is interesting from many points of view. The electromagnetic field, particularly in the optical region, is due to a large number of radiators, and a statistical approach is generally useful to describe the various physical effects. Although Maxwell's equations form the basis for the description of the various phenomena associated with radiation, the variation of the field characteristics cannot be measured with fine accuracy, and generally statistical averages of appropriate functionals play an important role. Curiously, the very process of detection introduces randomness due to its quantum nature. The process of detection consists in making the light-beam shine on a collection of photosensitive atoms which in turn shed electrons which are counted by a counter. Even if we consider a coherent beam of light, there are two sources of randomness: (i) the quantum nature of the electrons that are bound to the atoms, and (ii) the process of counting. Thus the number of electrons ejected over an interval of length T obeys a Poisson distribution with parameter proportional to T. The counter that counts the number of ejections is limited by its resolving time; the resolving time may be reduced but not absolutely to zero. This changes drastically the nature of the distribution viewed from a mathematical angle. There are more intricate situations wherein fluctuations of the field characteristics in general produce other types of phenomena whose characteristics can be quantified and perhaps measured. Thus a comprehensive stochastic modelling is necessary over and above the usual quantum description of the basic processes.

Viewed fundamentally, the phenomenon of cavity radiation is a consequence of the resonant interaction of the electromagnetic field with a micro-cavity endowed with various types of currents. The currents could be due to an atomic beam that is passed through it, or due to inclusions of solids containing magnetic dipole moments, or due to the dissipation of the field produced in the cavity wall. In other words, we have the electromagnetic field coupled to the material substance which carries the charge or the dipole moment. The problem can be studied from many angles; there is in fact a long history starting from Einstein's theory of radiation. Einstein postulated the theory of stimulated emission from general considerations of equilibrium reached thermally. The idea that stimulated emission could be used to amplify radiation gained currency much later. In fact, the main motivation for the study of amplification sprang from the advent of molecular amplifiers and modern detectors. The first maser device was constructed in 1954 by Gordan, Zeiger and Townes (1954); Basov and Prokhorov (1954) and Bloembergen (1956, 1965) suggested the use of multilevel atomic systems as the basis for a continuously operating solid-state maser. A theoretical model based on population growth was proposed by Shimoda, Takahasi and Townes (1957) who showed that many of the properties of amplifiers can be deduced from the birth, death and immigration model of a population of photons. About the same time Hanbury-Brown and Twiss (1954) measured the intensity correlations of light from a star in receivers that were separated, and showed how the correlations could be used to determine the angular diameter of stellar objects. This provided further stimulus for the study of coherent and partially coherent light-beams and various statistical properties associated with them. The practical significance of the intensity correlation brought to the fore the interesting problems arising from the detection of radiation in general and the counting process of photoelectrons in particular.

Speaking from a general philosophical angle of quantum logic, radiation and detection go together since measurements being themselves physical processes disturb the system. In view of the practical importance of the problem, there have been several parallel attempts resulting in different approaches to the question. Initially the intensity of radiation played an important role, and early attempts resulted in a semi-classical theory; the work of Lamb (1964), Mandel and Wolf (1970), and Haken (1970) are in this direction. Soon the advantage of using a fully quantum mechanical theory was realized, notably by Glauber (1963), Scully and Lamb (1967), Haken (1970), Louisell (1973) and many others (see for example Mandel and Wolf (1970)). The work of Feynman and Vernon (1963) and Schwinger (1960) also deserve special mention. There are again three distinct lines of approach. In the first, put forward by Scully and Lamb (1967, 1968, 1969), the density matrix is employed for the description of the cavity and field. An appropriate trace operation eliminates the co-ordinates of the cavity atoms, yielding ultimately an evolution equation for the field (expressed) in the number representation. The second line of approach is through the study of appropriate Green's functions called coherence functions; this was facilitated by the use of the coherent state representation or *P*-representation due to Glauber (1963) and Sudarshan (1963). The quantum detection process is characterized by an appropriate characteristic functional generated by the coherence functions. The third line of approach deals directly with the equations of motion for the second quantized operators; such attempts have been mainly by Haken (1970), Lax and Louisell (1967) and very often have led to viable Fokker–Planck equations; of course there is an added advantage in that the operators satisfy easily interpretable dynamical equations. In this monograph we will deal with some of the aspects of the evolution equation. As we have mentioned earlier, there are many elements that arise in an approach admittedly statistical; there is an added element of coarse-graining and time-scales that are adopted. These in turn introduce further statistical elements into the theory. The main object is the population of photons when it is in interaction with the cavity atoms and the detector, and we attempt to study the same by modelling the evolution as a population branching process.

The layout of the monograph is as follows. In Chapter 2 we summarize the main results of stochastic processes with special reference to Markov and population processes. The theory of point processes is presented in Chapter 3 with emphasis on population point processes and Cox processes. The presentation in these two chapters has a tutorial flavour and examples are discussed in detail. Then we discuss population growth problems mainly from the point of view of mathematical tools, with emphasis on the use of backward equation and regenerative character of the process in general. Chapter 5 introduces the physics of the problem and its various facets. Markov evolution is taken as the main theme of the next chapter; the method of analysis is rather new and has not appeared in the literature, although most of the results are well known. Non-Markov evolution is discussed in Chapter 7 where the population process evolves mainly as a Bellman–Harris type of process supported by an immigration process. Chapter 8 discusses modelling in terms of a Kendall type of age-dependent evolution. The division into two distinct types of evolution is only an analytical device and the physical process can be interpreted in terms of either type of evolution. Non-Markov evolution is shown to possess viability to bring out features like chaotic character, with versatile spectral profiles on the one hand and non-classical features like anti-bunching and sub-Poissonian behaviour on the other. So we proceed onward to the elements of stochastic theory!

2. Elements of stochastic processes

In this chapter we present some of the basic concepts of stochastic processes that are necessary for a proper understanding of the book. In doing so we do not attempt at either generality or completeness. The reader is assumed to have had an exposure to probability theory at the level of Parzen (1960) or Gnedenko (1962). We first define stochastic processes and summarize some of the main results relating to second-order processes; Gaussian processes are then introduced. We proceed to introduce the notion of Markov property and its consequences, particularly the Chapman–Kolmogorov relation. Some examples are then presented in detail to explain the general technique of obtaining explicit solution of Chapman–Kolmogorov equations in some special cases. Finally, multiplicative processes are introduced and the backward differential equation method is explained and some illustrative examples provided, with full working.

2.1 Stochastic processes

We start with a probability space $(\Omega, \mathbf{B}, P)$ and note that a random variable is just a real- or complex-valued measurable function $X(\omega)$ defined on $(\Omega, \mathbf{B})$. The notion of random variable is easily extended to vector- or function-valued random variables. In the case of function-valued random variable, we denote it explicitly by $X(\omega, t)$ ($t \in T$, $\omega \in \Omega$), where T is an arbitrary set. The case when t is the time parameter (or in general when T is an ordered parameter set) corresponds to the intuitive notion of a collection of random variables evolving with respect to t. It can be viewed as a mathematical model of a random phenomenon fluctuating with reference to the time parameter. We can define a stochastic process as follows:

Definition: Let $(\Omega, \mathbf{B}, P)$ be a probability space; T an arbitrary set of ordered elements, and $\mathcal{T}$ a σ-field of subsets of T. A *stochastic process* is a real (complex) valued function defined on $\Omega X T$, the mapping defined by

$$(\omega, t) \mapsto X(\omega, t)$$

being measurable with respect to the product σ-field $\mathbf{B} \times \mathscr{T}$.

Thus there are two ways of visualizing an arbitrary stochastic process $\{X(\omega, t), \omega \in \Omega, t \in T\}$:

(i) For each choice of $t \in T$, $X(., t)$ is a random variable, so that the stochastic process is a mapping of T into a set of real (complex) valued random variables.

(ii) For each choice of $\omega \in \Omega$, $X(\omega, .)$ is a real (complex) valued function and hence the stochastic process can be viewed as a mapping of Ω into the set of all possible functions or paths. The function $X(\omega, .)$ is called a *sample function* or *sample path*.

The set T is generally called the *index set* of the stochastic process and the set $\{\mathscr{R}_t, t \in T\}$, where $\mathscr{R}_t$ is the range of $X(., t)$, is called the *state space* of the process. This terminology is due to the influence of models of physical systems, and the state space symbolizes the set of values necessary to characterize the physical system in question. A stochastic process can be called a *discrete* or *continuous process* according as the state space is discrete or continuous. A similar dichotomy exists corresponding to the dichotomy of the index set T. A real stochastic process is specified completely by the family of finite dimensional distribution functions

$$F_n(x_1, t_1; x_2, t_2; \cdots x_n, t_n)$$

$$2.1.1 \qquad = Pr\{\omega : x(\omega, t_1) \leqslant x_1, \quad X(\omega, t_2) \leqslant x_2, \ldots, X(\omega, t_n) \leqslant x_n\}$$

for all possible values of t_i, x_i $(i = 1, 2, \ldots, n)$ and $n = 1, 2, \ldots$.

Stationary processes

We take the index set T to be reals. A stochastic process $\{X(t); t \in T\}$ is *strictly stationary*, or simply *stationary*, if its finite dimensional distributions are invariant under arbitrary translation of the index parameter:

$$F_n(x_1, t_1 + h; x_2, t_2 + h, \ldots; x_n, t_n + h)$$

$$2.1.2 \qquad = F_n(x_1, t_1; x_2, t_2; \ldots; x_n, t_n)$$

$$\text{for } \forall h, t_i \in T \ (i = 1, 2, \ldots, n) \quad n = 1, 2, \ldots.$$

If $X(t)$ for each t has a second moment and if $E[X(t)]$ and $\text{cov}[X(t+h), X(t)]$ are independent of t, then the stochastic process is called a weakly stationary process and $\gamma(h)$ defined by

2.1.3 $\qquad \gamma(h) = \text{cov}[X(t+h), X(t)]$

is called the *covariance* of the process $X(t)$. The process is also known as a *covariant stationary process*.

Example 2.1.1 Let $\{X_i; i = 1, 2, \ldots\}$ be a family of independent and identically distributed (i.i.d.) real random variables. Then $\{X_i\}$ is a stochastic process with index set N, the set of natural numbers and the state space as reals. The process is strictly stationary; the process is also weakly stationary provided X_i has a second moment.

Example 2.1.2 Let U and V be two independent and identically distributed random variables with zero mean and finite variance σ^2. Consider the process $\{X(t): -\infty < t < \infty\}$ when

2.1.4 $\qquad X(t) = U \cos \lambda t + V \sin \lambda t,$

where λ is any real number. The state space is the reals and the process $\{X(t)\}$ is not stationary. However, it is weakly stationary and the covariance function $\gamma(.)$ is given by

2.1.5 $\qquad \gamma(t) = \sigma^2 \cos \lambda t.$

Example 2.1.3 Simple Poisson process Let $\{X_i: i = 1, 2, \ldots\}$ be a family of i.i.d. non-negative random variables with the common distribution

2.1.6 $\qquad Pr\{X_i \leqslant x\} = 1 - e^{-\lambda x} \quad (\lambda > 0).$

If we define $N(t)$ $(t \geqslant 0)$ *by*

2.1.7 $\qquad N(t) = \sup\{n: X_1 + X_2 + \cdots + X_n \leqslant t\}$

then the process $\{N(t); t \geqslant o\}$ is a non-negative integer-valued process. If we define Y_n by

2.1.8 $\qquad Y_n = X_1 + X_2 + \cdots + X_n,$

then we have

2.1.9 $\qquad Pr\{N(t) = 0\} = Pr\{Y_1 > t\} = e^{-\lambda t}$

$$Pr\{N(t) = m\} = Pr\{Y_m < t, Y_{m+1} > t\} \quad (m > 0)$$

$$= e^{-\lambda t} \frac{(\lambda t)^m}{m!}.$$

In other words the random variable $N(t)$ for fixed t obeys a Poisson distribution with parameter λt. The process $\{N(t):t \geqslant 0\}$ is called a Simple Poisson process. The process has the interesting property that for $0 < t_1 < t_2$, the random variable $N(t_2) - N(t_1)$ has a Poisson distribution with parameter $\lambda(t_2 - t_1)$. It can be easily proved that the process has *stationary independent increments* in the sense that for all choices of $0 \leqslant t_0 < t_1 < t_2 < \cdots < t_n$ the random variables

$$N(t_1) - N(t_0), N(t_2) - N(t_1), \ldots, N(t_n) - N(t_{n-1})$$

form a family of independent random variables with the further property that $N(t_1 + h) - N(t_0 + h)$ has the same distribution as $N(t_1) - N(t_0)$ for all $t_0 < t_1$ and $h > 0$.

Gaussian (normal) process

A real-valued stochastic process $\{X(t)\}$ is called a *Gaussian process* or a *normal* process if $\{X(t_i):i = 1, 2, \ldots, n\}$ is jointly normally distributed for every choice of n and $\{t_i:i = 1, 2, \ldots, n\}$. Thus a Gaussian stochastic process is uniquely characterized by the expected value function $m(t) = E[X(t)]$ and the covariance function $R(t, s)$ defined by

$$2.1.10 \qquad R(t, s) = E\{[X(t) - m(t)][X(s) - m(s)]\}.$$

It is noted that a strictly stationary Gaussian process is also covariant stationary; the converse is also true since the characteristics of the process are determined only by $m(t)$ and $R(t, s)$. Thus the terminology "stationary Gaussian process" is unambiguous and we have in such a case

$$2.1.11 \qquad m(t) = \text{a constant}$$

$$2.1.12 \qquad R(t, s) = \text{a function of } |t - s|.$$

A *Complex Gaussian process* is defined to be a process whose real and imaginary parts are independent and identically distributed normal processes. It is to be specially noted that this definition requires a stronger condition than the requirement of normal property for the joint distribution of real and imaginary parts. Such a definition is necessary to make it possess properties similar to those of the real Gaussian process. In this case the covariance function $R(t, s)$ is defined by

$$2.1.13 \qquad R(t, s) = E\{[X(t) - m(t)][\overline{X(s)} - \overline{m(s)}]\}.$$

Karuhnen–Loeve expansion

We now consider complex-valued processes that are not necessarily normal and we take T to be the closed interval $[a, b]$. We further assume $E[|X(t)|^2] < \infty$ for every t. Such processes are called second-order processes. The covariance function $R(t, s)$ has the following properties:

$$2.1.14 \qquad R(t, s) = R(s, t), \quad R(t, t) \geqslant 0,$$

$$2.1.15 \qquad |R(t, s)| \leqslant \sqrt{R(t, t)R(s, s)};$$

if $\lambda_1, \lambda_2, \ldots, \lambda_n$ are complex numbers, then for any set of points $t_1, t_2, \ldots, t_n$ and any value of n we have

$$2.1.16 \qquad \sum_{j,k=1}^{n} R(t_j, t_k)\lambda_j \bar{\lambda}_k \geqslant 0.$$

When the above property holds good, we say that the function $R(t, s)$ is non-negative definite. We state without proof the following well-known theorem in analysis (see for example Riesz and Nagy (1955)).

Mercer's theorem: Let $R(t, s)$ be non-negative definite and continuous in $T \times T$. Then $R(t, s)$ has the representation

$$2.1.17 \qquad R(t, s) = \Sigma_n \lambda_n \phi_n(t) \bar{\phi}_n(s),$$

where the series on the r.h.s. converges absolutely and uniformly in both the variables and $\{\phi_n(t): n = 1, 2, \ldots\}$ is an orthonormal set of functions over the interval $[a, b]$ and $\phi_n(t)$ is determined by the eigen-vector of the integral equation

$$2.1.18 \qquad \int_a^b R(t, s)\phi(s)\mathrm{d}s = \lambda\phi(t) \quad a \leqslant t \leqslant b$$

corresponding to the eigen-value λ_n.

Now we are in a position to state without proof the *Karuhnen–Loeve representation theorem* Let $\{X(t): -\infty < a \leqslant t \leqslant b < \infty\}$ be a quadratic mean continuous second-order process with covariance function $R(t, s)$. Then the process has the representation

$$2.1.19 \qquad X'(t) = \Sigma_n \phi_n(t) Z_n,$$

where $\phi_n(t)$ $n = 1, 2, \ldots$ are the orthonormal eigen-functions of the integral equation (2.1.18) and $\{Z_n\}$ is a family of random variables with the property

2.1.20
$$E[Z_n] = 0$$

2.1.21
$$E[Z_n \bar{Z}_m] = \lambda_m \delta_{mn},$$

where λ_m is the eigen-value corresponding to the eigen-function $\phi_m(t)$, the equality 2.1.19 being defined in the quadratic mean sense. Moreover, if the process is Gaussian, $\{Z_n\}$ is a Gaussian process.

While the Karuhnen–Loeve representation is a very convenient one for practical use, the determination of the functions $\phi_n(t)$ is by no means an easy job. In view of this, we shall indicate the method in one instance which will be of use in the next chapter.

Example 2.1.4 Let $\{X(t): -a \leqslant t \leqslant a\}$ be a second-order process with $E[X(t)] = 0$ and covariance $R(t, s)$ given by

2.1.22
$$R(t, s) = e^{-\beta|t - s| + i\omega(t - s)} \quad (\beta > 0).$$

The eigen-functions are determined by

2.1.23
$$\int_{-a}^{a} e^{-\beta|t - s| + i\omega(t - s)} \phi(s)ds = \lambda\phi(t).$$

It is convenient to define

2.1.24
$$f(s) = \phi(s)^{-i\omega s}$$

and rewrite 2.1.23 as

2.1.25
$$\int_{-a}^{a} e^{-\beta|u - v|} f(v)dv = \lambda f(u).$$

Introducing the Laplace transform (LT) $f^*(p)$ by

2.1.26
$$f^*(p) = \int_{-a}^{a} f(u)\, e^{-up} du$$

we obtain

2.1.27
$$f^*(p) = \frac{2\beta}{\beta^2 - p^2} f^*(p) - \frac{f^*(-\beta)e^{-(\beta + p)a}}{p + \beta}$$
$$+ \frac{f^*(\beta)e^{(p - \beta)a}}{p - \beta}.$$

Thus the solution for $f^*(p)$ is given by

2.1.28 $$f^*(p) = \frac{(p+\beta)f^*(\beta)\,e^{(p-\beta)a} - (p-\beta)f^*(-\beta)\,e^{-(p+\beta)a}}{\lambda(p^2-\beta^2)+2\beta}.$$

We note that $f^*(p)$ is an analytic function of p; however, an inspection of the r.h.s. of 2.1.28 shows that there are two poles corresponding to the zeros of the denominator given by

2.1.29 $$\lambda(p^2-\beta^2)+2\beta = 0.$$

Denoting the zeros by $\pm p_0$, we find

2.1.30 $$p_0^2 = \beta^2 - \frac{2\beta}{\lambda}.$$

The analytic nature of $f^*(p)$ then implies that the numerator of the r.h.s. vanishes for $p = \pm p_0$:

2.1.31 $$(p_0+\beta)f^*(\beta)\,e^{(p_0-\beta)a}$$
$$- (p_0-\beta)f^*(-\beta)\,e^{-(p_0+\beta)a} = 0.$$

2.1.32 $$(-p_0+\beta)f^*(\beta)\,e^{-(p_0+\beta)a}$$
$$+ (p_0+\beta)f^*(\beta)\,e^{+(p_0-\beta)a} = 0.$$

Eliminating $f^*(\beta)$ and $f^*(-\beta)$, we obtain

2.1.33 $$\left(\frac{p_0+\beta}{p_0-\beta}\right) = \pm e^{-2p_0 a}$$

which can be interpreted to be the equation giving the eigen-values λ. Putting $p_0 = i\omega$ and reducing equation 2.1.33 further, we finally obtain the twin equations

2.1.34 $$\omega \tan \omega a = \beta, \qquad \omega \cot \omega a = -\beta,$$

which prove the number of eigen-values λ satisfying 2.1.34 are *countable*.

Using the condition 2.1.33 in 2.1.31 or 2.1.32, we find

2.1.35 $$f^*(p) = \frac{f^*(\beta)\,e^{-\beta a}}{\lambda(p^2-(\beta^2))+2\beta}\left[(p+\beta)\,e^{pa} + (p-\beta)\,e^{-pa}\right]$$

when the $+$ sign in 2.1.33 is chosen. On the other hand, when the $-$ sign is chosen, we have

2.1.36 $$f^*(p) = \frac{f^*(\beta)\,e^{-\beta a}}{\lambda(p^2-\beta^2)+2\beta}\left[(p+\beta)\,e^{pa} - (p-\beta)\,e^{-pa}\right].$$

Observe that $f^*(\beta)$ is an undetermined constant in either choice; this only shows that the eigen-functions $f(t)$ are determined except for a constant multiplier. However this can be uniquely fixed by using the normalization condition

$$2.1.37 \qquad \int_{-a}^{a} |f(t)|^2 dt = 1.$$

Inverting the Laplace transform, we finally obtain

$$2.1.38 \qquad f(t) = \left(\frac{1}{a} \frac{1}{1 - \dfrac{\sin 2\omega a}{2\omega a}} \right)^{1/2} \sin \omega t$$

when the eigen-value is given by

$$2.1.39 \qquad \omega \cot \omega a = -\beta$$

and

$$2.1.40 \qquad f(t) = \left(\frac{1}{a} \frac{1}{1 + \dfrac{\sin 2\omega a}{2\omega a}} \right)^{1/2} \cos \omega t$$

when the eigen-value is given by

$$2.1.41 \qquad \omega \tan \omega a = \beta.$$

It is easily verified from 2.1.30 and 2.1.33 that the resulting eigen-values λ are positive. Thus we conclude that the random process $\{X(t)\}$ whose covariance function is given by 2.1.22 is a weighted sum of sine and cosine functions of ωt.

2.2 Markov processes

A stochastic process $\{X(t): t \in T\}$ is called a *Markov* process if for $s < t$, and given the value of $X(s)$, the probability distribution of $X(t)$ is independent of the values of $X(u)$, $u < s$. In other words, the conditional distribution of $X(t)$ given $X(u)$, for $u \leqslant s \leqslant t$, is independent of all values of $u < s$. Thus, more formally, the process $\{X(t), t \in T\}$ is a Markov process, if

$$Pr\{X(t) \leqslant x \mid X(t_1) = x_1, X(t_2) = x_2, \ldots, X(t_n) = x_n\}$$

$$2.2.1 \qquad = Pr\{X(t) \leqslant x \mid X(t_n) = x_n\}$$

whenever $t_1 < t_2 < \cdots < t$ $(n = 2, 3, \ldots)$. A Markov process is completely specified if the first-order distribution $F_1(x, t)$ and the conditional first-order distribution $F_2(x_2, t_2 \mid x_1, t_1)$ $(t_1 < t_2)$ defined by

$$2.2.2 \qquad F_2(x_2, t_2 \mid x_1, t_1) = Pr\{X(t_2) \leqslant x_2 \mid X(t_1) = x_1\}$$

are known. If the state space is discrete, then the Markov condition 2.2.1 can be written in a more elegant form. If

$$\pi_1(x_1, t_1), \pi_2(x_2, t_2 \mid x_1, t_1), \ldots,$$

$$\pi_n(x_n, t_n \mid x_1, t_1; x_2, t_2; \cdots x_{n-1}, t_{n-1})$$

are the corresponding probability mass functions so that

$$2.2.3 \qquad \pi_1(x, t_1) = Pr\{X(t_1) = x_1\}$$

$$2.2.4 \qquad \pi_2(x_2, t_2 \mid x_1, t_1) = Pr\{X(t_2) = x_2 \mid X(t_1) = x_1\}$$

$$\pi_n(x_n, t_n \mid x_1, t_1; x_2, t_2, \ldots, x_{n-1}, t_{n-1})$$

$$2.2.5 \qquad = Pr\{X(t_n) = x_n \mid X(t_1) = x_1,$$

$$x(t_2) = x_2 \cdots X(t_{n-1}) = x_{n-1}\}$$

then the Markov condition 2.2.1 takes the form

$$\pi_n(x_n, t_n \mid x_1, t_1; x_2, t_2 \cdots; x_{n-1}, t_{n-1})$$

$$2.2.6 \qquad = \pi_2(x_n, t_n \mid x_{n-1}, t_{n-1})$$

whenever $t_1 < t_2 < \cdots < t_n$.

Chapman–Kolmogorov equation

The Markov condition implies a non-trivial condition satisfied by the second-order probability mass function. To see the point, we consider the functions $\pi_2(x_2, t_2 \mid x_1, t_1)$. Let us assume $t_1 < t_2$; consider a point t' such that $t_1 < t' < t_2$. Then by the law of inclusive probabilities we have

$$\pi_2(x_2, t_2 \mid x_1, t_1)$$

$$2.2.7 \qquad = \sum_{x'} Pr\{X(t_2) = x_2, X(t') = x' \mid X(t_1) = x_1\}.$$

Using the law of multiplication of probabilities, we have

$$\pi_2(x_2, t_2 \mid x_1, t_1)$$

$$2.2.8 \qquad = \sum_{x'} \pi_3(x_2, t_2 \mid x_1, t_1, x', t') \pi_2(x', t' \mid x_1, t_1).$$

If at this stage we use the Markov constraint in the form 2.2.6 we obtain

$$\pi_2(x_2, t_2 \,|\, x_1, t_1)$$

2.2.9
$$= \sum_{x'} \pi_2(x_2, t_2 \,|\, x', t')\pi_2(x', t' \,|\, x_1, t_1)$$

a relation which holds good for arbitrary values of x_2 and x_1 and t_1, t', t_2 subject to $t_1 < t' < t_2$. Relation 2.2.9 is known as the *Chapman–Kolmogorov relation*. Since the state space is discrete, we can replace the x's by discrete indices and write

2.2.10
$$\pi_2(x_2, t_2 \,|\, x_1, t_1) = \rho_{ij}(t_2 \,|\, t_1)$$

so that 2.2.9 can be written in the elegant form

2.2.11
$$\rho_{ij}(t_2 \,|\, t_1) = \sum_k \rho_{ik}(t_2 \,|\, t')\rho_{kj}(t' \,|\, t_1).$$

If in addition the process is stationary, then ρ_{ij} is a function of $t_2 - t_1$ only so that 2.2.11 takes the form

2.2.12
$$\rho_{ij}(t + s) = \sum_k \rho_{ik}(t)\rho_{kj}(s) \quad (t, s > 0).$$

The function $\rho_{ij}(\,.\,)$ is known as the stationary transition probability. The above relation shows that the class of functions $\{\rho\}$ has the semi-group property which can be used to characterize the random process. An elementary account with illustrative examples is given by Feller (1968); for a more rigorous treatment, we refer the reader to Feller (1971) and Chung (1967).

The above considerations are equally valid when the state space is continuous. The Chapman–Kolmogorov relation 2.2.11 is valid provided the process admits probability densities, in which case we interpret $\pi_2(x_2, t_2 \,|\, x_1, t_1)$ as the conditional density function of $X(t_2)$ given $X(t_1) = x_1$; the summation over x' will be replaced by integration over x'.

The Chapman–Kolmogorov relation is too general in character, and at best some weak structural relations that follow from the semi-group property can be deduced. To make further progress, we impose some constraints on the transition probability function. If we assume that the process is stationary and that

2.2.13
$$\rho_{ij}(\Delta) = R_{ij}\Delta + o(\Delta) \quad i \neq j$$

then by putting $s = \Delta$, 2.2.12 can be written in the form

2.2.14
$$p_{ij}(t + \Delta) = \sum_{k \neq j} p_{ik}(t) R_{kj} \Delta$$
$$+ \left(1 - \sum_{k \neq j} R_{kj} \Delta \right) p_{ij}(t) + o(\Delta).$$

Proceeding to the limit as $\Delta \to 0$, we obtain

2.2.15
$$p'_{ij}(t) = \sum_{\substack{k \\ k \neq j}} p_{ik}(t) R_{kj} - \left(\sum_{\substack{k \\ k \neq j}} R_{kj} \right) p_{ij}(t).$$

If on the other hand we put $t = \Delta$ in 2.2.12 and proceed to the limit as $\Delta \to 0$, we obtain

2.2.16
$$p'_{ij}(s) = \sum_{\substack{k \\ k \neq i}} p_{kj}(s) R_{ik} - \left(\sum_{\substack{k \\ k \neq i}} R_{ki} \right) p_{ij}(s).$$

The elements R_{kj} are called *stationary transition rates* $(j \to k)$ and the equations 2.2.15 and 2.2.16 are known respectively as Kolmogorov backward and Kolmogorov forward equations. Note that in 2.2.15 the symbol i remains fixed while in 2.2.16 j remains fixed. The two systems of equations can be seen to be equivalent by writing them in matrix notation. However, the two systems of equations are not equivalent if we relax the stationarity condition. There are some situations where the forward system does not even exist.

Example 2.2.1   Simple Poisson Process   Let the state space be the set of non-negative integers and the index set, the real line. Let the Markov Process be stationary. Define the stationary transition rates by

2.2.17
$$R_{jk} = \lambda \quad j = k + 1$$
$$= 0 \quad \text{otherwise.}$$

The process $\{X(t)\}$ is a pure additive process in the sense

2.2.18
$$X(t_2) = X(t_1) + X(t_2 - t_1) \quad (t_2 > t_1)$$

with probability one. The probability $p_{ij}(t_2 | t_1)$ defined by 2.2.12 is a function of $i - j$ and $t_2 - t_1$ only; accordingly we write

2.2.19
$$p_{ij}(t_2 | t_1) = \pi_{i-j}(t_2 - t_1).$$

Thus it is convenient to introduce a counting process $\{N(t); t \geqslant 0\}$ where

2.2.20
$$N(t) = X(t_1 + t) - X(t_1) \quad (t_1 \in R).$$

Then it follows that

2.2.21 $\qquad Pr\{N(t) = m\} = \pi_m(t).$

The forward and backward equations become identical:

2.2.22 $\qquad \pi'_m(t) = -\lambda\pi_m(t) + \lambda\pi_{m-1}(t) \quad (m \geqslant 1.$

$$\pi'_0(t) = -\lambda\pi_0(t).$$

Using the initial condition

2.2.23 $\qquad \pi_m(0) = \delta_{m0}$

we obtain

2.2.24 $\qquad \pi_m(t) = e^{-\lambda t}(\lambda t)^m/m! \quad m = 0, 1, 2, \ldots$

To sum up: the counting process $\{N(t); t \geqslant 0\}$ defined by 2.2.20 has the following properties:

(i) $N(0) = 0$ with probability 1;

(ii) $\{N(t); t \geqslant 0\}$ has stationary independent increments (by virtue of stationarity of the process $X(t)$ and 2.2.17 and 2.2.20);

(iii) $Pr\{N(\Delta) = 1\} = \lambda\Delta + o(\Delta)$

(iv) $Pr\{N(\Delta) > 1\} = o(\Delta)$

$\left.\right\}$ by virtue of 2.2.17.

Example 2.2.2.   Two-state process   Let the state space consist of two points $\{0, 1\}$ and the index set the real line. Let the process which we denote by $\{X(t): t \in R\}$ be a stationary Markov process. There are only two stationary transition rates R_{10} and R_{01}. Denote them by α and β. Stationarity implies that

2.2.25 $\qquad \pi_i(t) = Pr\{X(t) = i\}$

$$= \text{constant} = \pi_i \quad i = 0, 1.$$

We determine the constants by invoking the Markov property (Chapman–Kolmogorov relation):

2.2.26 $\qquad \pi_0(t + \Delta) = \pi_0(t)(1 - \alpha\Delta) + \pi_1(t)\beta\Delta + o(\Delta)$

Using 2.2.25 and the fact that $\pi_0 + \pi_1 = 1$, we obtain

2.2.27 $\qquad \pi_0 = \pi_0(1 - \alpha\Delta) + (1 - \pi_0)\beta\Delta,$

2.2.28 $\qquad \pi_0 = \beta/(\alpha + \beta),$

2.2.29 $\qquad \pi_1 = \alpha/(\alpha + \beta).$

Next we write down the Kolmogorov forward differential equation for $\rho_{00}(t)$:

2.2.30 $\qquad \rho'_{00}(t) = -\alpha\rho_{00}(t) + \beta\rho_{10}(t).$

Using the relation

2.2.31 $\qquad \rho_{00}(t) + \rho_{10}(t) = 1$

we obtain

2.2.32 $\qquad \rho_{00}(t) = [\beta + \alpha e^{-(\alpha+\beta)t}]/(\alpha+\beta),$

2.2.33 $\qquad \rho_{10}(t) = \alpha[1 - e^{-(\alpha+\beta)t}]/(\alpha+\beta).$

In a similar way we obtain

2.2.34 $\qquad \rho_{11}(t) = [\alpha + \beta e^{-(\alpha+\beta)t}]/(\alpha+\beta)$

2.2.35 $\qquad \rho_{01}(t) = \beta[1 - e^{-(\alpha+\beta)t}]/(\alpha+\beta).$

The functions $\rho_{ij}(.)$ and the constants π_0 and π_1 completely characterize the process.

Example 2.2.3   Random telegraph signal   In Example 2.2.2 we relabel the points 0, 1 of the state space by -1 and 1. The process is characterized by the functions $\rho_{ij}(.)$ except for change of notation. If we set $\alpha = \beta$, we obtain the *Random telegraph signal*. The covariance function $\gamma(.)$ is easily evaluated:

2.2.36 $\qquad \gamma(t) = e^{-2\alpha|t|}.$

If we denote $Y(t_0, t_1)$ (for $t_0 < t_1$) the number of transitions experienced by the process over the interval (t_0, t_1), then $Y(t_0, t_1)$ is a non-negative integer-valued (random) function of the argument $t_1 - t_0$ and is just a counting process. If we denote it by $\{N(t); t \geq 0\}$, it is easy to establish that it is a *simple Poisson process*.

Example 2.2.4   Counting process of visits   Let us consider Example 2.2.2 and denote by $Y_i(t_0, t_1)$ $(t_0 < t_1)$ the number of transitions to state i $(i = 0, 1)$ over the interval (t_0, t_1). Then $Y(t_0, t_1)$ is a counting process with argument $t_1 - t_0$. If we set $t_0 = 0$ and denote the resulting counting process by $\{N_i(t); t > 0\}$ we note that the conditional probability

$\qquad Pr\{N_i(t)\,|\,N_i(t')\} \quad (t' < t)$

is not unique and depends on the evolution of the process $X(.)$ over the interval $[0, t']$. Consequently the process $\{N_i(t)\}$ is not Markov; on the other hand, if we consider the two-dimensional process $\{N_i(t), X(t)\}$, then the new process has Markov property. It is convenient to use the backward differential equation technique to determine the probabilistic structure of the process. Introducing the probabilities $\pi_j^i(n, t)$,

2.2.37 $\qquad \pi_j^i(n, t) = Pr\{N_i(t) = n \,|\, X(0) = j\} \quad i, j = 0, 1,$

we obtain

$$\frac{d}{dt} \pi_0^0(n, t) = -\alpha \pi_0^0(n, t) + \alpha \pi_1^0(n, t)$$

2.2.38 $\qquad \dfrac{d}{dt} \pi_1^0(n, t) = -\beta \pi_1^0(n, t) + \beta \pi_0^0(n - 1, t)$

with the initial conditions

2.2.39 $\qquad \pi_0^0(n, 0) = \pi_1^0(n, 0) = \delta_{n0}.$

The probabilities $\pi_i^0(n, t)$ can be obtained by solving the differential equations by the combined use of generating function and Laplace transform technique.

Example 2.2.5 Birth-and-death process Let the state space be the set of non-negative integers and the index set, positive reals. Define the Markov transition probabilities by

2.2.40 $\qquad \pi_2(n, t + \Delta \,|\, m, t) = \begin{cases} m\lambda\Delta + o(\Delta) & m = n - 1 \\ m\mu\Delta + o(\Delta) & m = n + 1 \\ 1 - m(\lambda + \mu)\Delta + o(\Delta) & m = n \\ 0 & \text{otherwise,} \end{cases}$

where λ and μ are non-negative constants. The Kolmogorov forward differential equation is given by

$$\frac{d}{dt} \pi_2(n, t \,|\, m, t) = -n(\lambda + \mu)\pi_2(n, t \,|\, m, t_0)$$

$$+ (n - 1)\lambda \pi_2(n - 1, t \,|\, m, t_0)$$

$$+ (n + 1)\mu \pi_2(n + 1, t \,|\, m, t_0) \quad n \geqslant 1$$

2.2.41 $\qquad \dfrac{d}{dt} \pi_2(0, t \,|\, m, t_0) = \mu \pi_2(1, t \,|\, m, t_0)$

with the initial condition

2.2.42 $\qquad \pi_2(n, t_0 | m, t_0) = \delta_{nm}.$

Equations 2.2.41 can be solved by the use of the generating function defined by

2.2.43 $\qquad g_2(z, t | m, t_0) = \sum z^n \pi_2(n, t | m, t_0).$

Thus, multiplying 2.2.41 by z^n and summing over all values of n, we obtain

$$\frac{\partial}{\partial t} g(z, t | m, t_0)$$

2.2.44 $\qquad = \frac{\partial}{\partial z} g(z, t | m, t_0)(z - 1)(\lambda z - \mu) \quad (t > t_0)$

with the boundary condition

2.2.45 $\qquad g(z, t_0 | m, t_0) = z^m.$

From the solution of the above partial differential equation we can obtain the probabilities we started with.

There is an interesting way to simplify the analysis. This comes from an interpretation of the Markov transition probabilities 2.2.40. If we take the process $\{X(t)\}$ to represent a population, then $X(t)$ denotes the number of members of the population. If $\lambda\Delta + o(\Delta)$ is the probability that a member produces a replica in the time-interval $(t, t + \Delta)$ and likewise if $\mu\Delta + o(\Delta)$ is the probability of death of a member of the population in the time-interval $(t, t + \Delta)$, then 2.2.40 implies that the members of the population evolve independently, since

$$m\lambda\Delta = \sum \lambda\Delta$$

$$m\mu\Delta = \sum \mu\Delta$$

2.2.46 $\qquad 1 - m(\lambda + \mu)\Delta = \prod (1 - (\lambda + \mu)\Delta),$

where the summation and product is over the different individuals in the population. This implies that the transition probability function $\pi_2(., t | m, t_0)$ can itself be obtained by an m-fold convolute of $\pi_2(., t | 1, t_0)$. In terms of generating functions, we have the relation

2.2.47 $\qquad g(z, t | m, t_0) = [g(z, t | 1, t_0)]^m.$

Thus it is sufficient if we deal with $\pi_2(., t | 1, t_0)$ or $g(z, t | 1, t_0)$. If at this stage, we use the backward differential equation instead of the forward one, the analysis can be further simplified. Before we do this we note that the transition probabilities are independent of t and hence t-dependence of the function $\pi_2(n, t \mid 1, t_0)$ is only through $t - t_0$. Let us set $t_0 = 0$ without loss of generality and denote $\pi_2(n, t \mid 1, 0)$ by simply $\pi_2(n, t \mid 1)$. Next we write the Chapman–Kolmogorov relation when the intermediate point is close to the origin. Thus we have

$$2.2.48 \qquad \pi_2(n, t \mid 1) = \sum_m \pi_2(m, \Delta \mid 1)\pi_2(n, t - \Delta \mid m)$$

which can be recast in terms of the generating function by using 2.2.47:

$$2.2.49 \qquad g(z, t \mid 1) = \sum_m \pi_n(m, \Delta \mid 1) \ [g(z, t - \Delta \mid 1)]^m.$$

Observing that

$$2.2.50 \qquad \pi_2(m, \Delta \mid 1) = \{1 - (\lambda + \mu)\Delta\}\delta_{m1} + \lambda\Delta\delta_{m2} + \mu\Delta\delta_{m0} + o(\Delta),$$

we obtain by proceeding to the limit, as $\Delta \to 0$,

$$\frac{\partial g(z, t \mid 1)}{\partial t} = -(\lambda + \mu)g(z, t \mid 1)$$

$$2.2.51 \qquad\qquad\qquad + \lambda[g(z, t \mid 1)]^2 + \mu$$

with the initial condition

$$2.2.52 \qquad g(z, 0 \mid 1) = z.$$

Conceptually, 2.2.51 is simpler in structure and this is the advantage of the backward differential equation. The solution of 2.2.51 is easily obtained:

$$2.2.53 \qquad g(z, t \mid 1) = 1 - \frac{(\lambda - \mu)(z - 1)\,e^{-(\mu - \lambda)t}}{\mu - \lambda z + \lambda(z - 1)\,e^{-(\mu - \lambda)t}}.$$

If $\mu > \lambda$, $g(z, t \mid 1)$ tends to 1 as t tends to infinity, showing that the population becomes extinct with probability 1 as t tends to infinity. On the other hand, if $\mu < \lambda$, $g(z, t \mid 1)$ tends to $\mu/\lambda < 1$, showing that the population has a mass concentration at infinity. The first few moments are easily obtained by appropriate differentiation and are given by

$$2.2.54 \qquad E[X(t)] = e^{-(\mu - \lambda)t}$$

$$2.2.55 \qquad E\{[X(t)]^2\} = E[X(t)] + \frac{2\lambda}{\mu - \lambda}\,e^{-(\mu - \lambda)t}(1 - e^{-(\mu - \lambda)t}).$$

A stochastic process $\{X(t)\}$ of the type discussed above is called a *multiplicative process* or more generally a *branching process*. The process, besides being discrete valued, has the characteristic property of being identified as a population, with the members evolving independently. The parameter t can be discrete or continuous. If t is discrete, the transition probability can be denoted by $\pi_{ij}(m, n)$, where

$$2.2.56 \qquad \pi_{ij}(m, n) = Pr\{X_i = m \mid X_j = n\} \quad i > j.$$

In this case we say that the process is a branching process if

$$2.2.57 \qquad \pi_{ij}(m, n) = \pi_{ij}^{(n)}(m, 1),$$

where $\pi_{ij}^{(n)}(., 1)$ is an n-fold convolute of $\pi_{ij}(., 1)$. Such processes are also known as *multiplicative chains*. Example 2.2.5 is a typical Markov branching process. Non-Markov branching processes arise when the evolution is dependent on some specific characteristic of the individual. For instance the birth-and-death rates could be age-specific. Such processes are generally known as age-dependent branching processes, and special techniques are necessary to deal with them. We will deal with some specific types of age-dependence in Chapter 4.

Finally we consider stationary Gaussian processes. Such processes are characterized by the covariance function. There is a theorem due to Doob which characterizes Markov processes from among stationary Gaussian processes. It simply states that *the covariance function of a Gaussian process is an exponential function of $k(t - s)$ (where k is some complex constant) if and only if the process is Markov.* Thus for a stationary complex Gaussian process, the Markov property implies

$$2.2.58 \qquad R(t, s) = e^{-a|t - s| + ib(t - s)}.$$

Thus Example 2.1.4 provides a kind of Fourier expansion of such a process.

So far we have dealt with discrete-valued process in great detail by examples. We conclude the chapter by an example in which the state space is a continuum, the process making jumps with the characteristic property that the number of jumps in any finite interval of time remains finite, the process itself remaining constant between two such jumps with probability 1. Such processes are known as *purely discontinuous processes* or *Feller–Kolmogorov processes*.

Example 2.2.6 Energy straggling of fast particles A high-energy charged particle in its passage through matter radiates quanta and loses energy. We can model such a phenomenon by a stochastic process with state space forming a continuum. The loss of energy by the emission of a photon is described by the quantum mechanical cross-section. More specifically, a particle of energy E has a probability $R_1(E, E')\mathrm{d}E'$ of dropping to an energy in the interval $(E', E' + \mathrm{d}E')$ while traversing matter of unit thickness—where $R_1(E, E')$ is given by

$$2.2.59 \qquad R_1(E, E') = \frac{\left\{\dfrac{4}{3}(E^2 - EE') + E'^2\right\}}{\left\{\left[1 + \omega^2\left(\dfrac{E}{E'}\right)^2\right]E^2 E'\right\}} \qquad E' \leqslant E,$$

where ω is a constant depending on the polarization characteristic of the material traversed. The value of ω is 1.9×10^{-4} and 7.5×10^{-5} (in radiation units) for air and lead respectively. We are interested in the probability distribution of the energy of the particle when it has traversed t units of matter (measured in radiation units), given the initial energy E_0. We treat the problem as one-dimensional so far as its penetration through matter is concerned. The energy state is described by the stochastic process $\{X(t) \quad t \geqslant 0\}$ with the state space as the interval $[0, E_0]$. The process is indeed Markov; if we introduce the transition probability density function $\pi(E, t \,|\, E_0)$, where

$$2.2.60 \qquad \pi(E, t \,|\, E_0) = \lim_{\Delta \to 0} Pr\{E < X(t) < E + \Delta \,|\, X(0) = E_0\}/\Delta,$$

then we can write the Chapman–Kolmogorov relation by choosing the parametric values $0, t, t + \Delta$. Observing that the probabilities depend only on the amount of thickness traversed, we obtain

$$\pi(E, t + \Delta \,|\, E_0) = \Delta \int_{E'} \pi(E', t \,|\, E_0) R_1(E', E)\mathrm{d}E'$$

$$2.2.61 \qquad + \pi(E, t \,|\, E_0)\{1 - \Delta \textstyle\int R_1(E, E')\mathrm{d}E'\} + o(\Delta).$$

By taking the limit as $\Delta \to 0$, we finally have the integro-differential equation

$$\frac{\partial \pi(E, t \,|\, E_0)}{\partial t} = -\pi(E, t \,|\, E_0) \int_0^{E_0} R_1(E, E')\mathrm{d}E'$$

$$2.2.62 \qquad + \int_E^{E_0} \pi(E', t \,|\, E_0) R_1(E', E)\mathrm{d}E'.$$

Next we note that the probability $R_1(E, E')\mathrm{d}E'$ depends only on the ratio

$x = E'/E$, and introducing $W(x)$ by

2.2.63 $\qquad R_1(E, E')\mathrm{d}E' = W(x)\mathrm{d}x,$

we identify $W(x)$:

2.2.64 $\qquad W(x) = \left[\frac{4}{3}\left(\frac{1}{x} - 1\right) + x\right]\bigg/\left[1 + \frac{\omega^2}{x^2}\right].$

Observing that $\pi(E, t\,|\,E_0)$ is a function of $\varepsilon = E/E_0$ and not of E and E_0 individually, and writing

2.2.65 $\qquad \pi(E, t\,|\,E_0) = \pi(\varepsilon, t) \quad (0 \leqslant \varepsilon \leqslant 1),$

we can rewrite the Kolmogorov forward differential equation 2.2.62 as

2.2.66 $\qquad \dfrac{\partial \pi(\varepsilon, t)}{\partial t} = -\pi(\varepsilon, t)\int_0^1 W(x)\mathrm{d}x + \int_\varepsilon^1 \pi\left(\dfrac{\varepsilon}{x}, t\right)\dfrac{W(x)}{x}\mathrm{d}x,$

an equation which is amenable to the Mellin transform technique. Defining

2.2.67 $\qquad \Gamma(s, t) = \int_0^1 \pi(\varepsilon, t)\varepsilon^{s-1}\mathrm{d}\varepsilon \quad Rls > 0,$

we obtain

2.2.68 $\qquad \dfrac{\partial}{\partial t}\Gamma(s, t) = -\Gamma(s, t)A(s),$

where

2.2.69 $\qquad A(s) = \int_0^1 W(x)(1 - x^s)\mathrm{d}x.$

Thus the solution is given by

2.2.70 $\qquad \Gamma(s, t) = \exp -tA(s),$

where we have made use of the fact that the initial condition corresponds to the probability mass concentration of unity at $\varepsilon = 1$. The probability density $\pi(\varepsilon, t)$ can be obtained by numerically inverting $\Gamma(s, t)$.

Additional notes

There are many books on stochastic processes and it is rather difficult to make any specific recommendation. Nevertheless, taking into account the interest of those who are likely to get access to this book, we recommend Bartlett (1975), Feller (1968), Karlin (1966), Parzen (1962) and Ramakrishnan (1958).

3. Point processes and product densities

We now present a summary of some of the main results of the theory of point processes. The theory of point processes has been developed in the last fifty years as a result of investigations of cosmic-ray cascades, statistical physics, population dynamics, response phenomena, and queuing and telephone traffic models. We present in this chapter mainly those results that are directly relevant to our modelling problem; however, we present additional historical notes at the end of the chapter which will provide a proper perspective of the development of the subject as a whole. No attempt will be made to derive many of the results in a rigorous manner. We first provide a proper setting for a working definition of a population point process mainly confined to the real line. Then we introduce the counting measures and the measures generated by the moments of the counting process; the product densities are then introduced in a natural way for the most general case. We then examine the implications of stationarity. The final section is devoted to a brief discussion on Cox processes. Since the concepts are fairly difficult, we have introduced a number of illustrative examples to facilitate better comprehension of the theory as a whole.

3.1 Population point processes

A point process is the mathematical abstraction that arises from the study of a special class of stochastic processes describing an abstract population whose members enjoy random characteristics. The random characteristics give rise to the set of points $\{x_n\}$ drawn from the state space X, $\{x_n\}$ representing the "locations" of the different members of the population. There are several interesting examples:

(1) The state space X is the real line, and the set of points $\{x_n\}$ from X may represent

(i) the collection of time-instants at which telephone calls arrive in an exchange or customers arrive in (depart from) a service facility, so that the population consists of the collection of calls arriving at the exchange at various instants of time or the collection of customers who arrive in (depart from) the service facility;

(ii) the energies of the cosmic-ray particles at fixed depth of the atmosphere, or the age of the various members of a real population at a fixed time.

(2) The state space X is the two-dimensional Euclidean plane, and the set of points $\{x_n\}$ from X may denote

(i) the location of points in a garden plot where the weeds grow;

(ii) the energy and depth of the cosmic-ray particles at the points of their production (as observed in emulsion experiments).

(3) The state space X is the surface of the sphere and $\{x_n\}$ represents the epicentres of earthquakes.

(4) The state space X is the Cartesian product of the real line R and a finite set Y. The set of points $\{x_n\}$ can then represent the energy and state of polarization of cosmic-ray electrons or muons at a definite depth of the atmosphere. If instead of real line R, we take the two-dimensional Euclidean plane, then we can take x_n to represent the energy, depth and state of polarization at the point of production of these particles as in (2) (ii) above.

In these examples, it is possible to associate a probability measure with each set of points $\{x_n\}$ from X and to study the random point distribution. Most of the initial work is in the context of the problems arising from the examples mentioned above, and hence the terminology point process has come to stay. The mathematical theory of point processes can be developed when X is a complete, separable σ-compact metric space. However, in the following we confine ourselves to the special case when X is a finite subset of the real line.

Let us consider a set of objects each of whose state is describable by a point x of a fixed set of points X. Such a collection of objects which we call a "population" is stochastic if there exists a well-defined probability distribution P on some σ-field $\mathbf{B}$ of subsets of the space Φ of all states. We shall assume that the members of the population are indistinguishable from each other. The state of the population is defined as an unordered set $x^n = \{x_1, x_2, \ldots, x_n\}$ representing the situation where the population has n members, one each in the states $x_1, x_2, \ldots, x_n$. Thus the population state space Φ is the collection of all x^n with $n = 0, 1, 2, \ldots$, and $x_i \in X$. We take x^0 to denote the empty population. A point process is the triplet $(\Phi, \mathbf{B}, P)$. It is possible to define measurable functions on Φ since $(\Phi, \mathbf{B})$ is a measure space.

There is an alternative way of characterizing the population by assigning to sets A in the state space X, the number of objects or individuals $N(A)$ each of whose state x belongs to A ($x \in A$). We fix our attention on the population whose total size is finite with probability one. To make further progress, we consider the special case when the population consists of n individuals in the states $x_1, x_2, \ldots, x_n$. The number of individuals with states in a given arbitrary $A \subset X$ is given by

$$3.1.1 \qquad N(A \mid x^n) = \sum_{i=1}^{n} \delta(A \mid x_i),$$

where $\delta(A \mid .)$ is the characteristic function of the set A and is defined by

$$3.1.2 \qquad \delta(A \mid x) = 1 \quad x \in A$$

$$= 0 \quad \text{otherwise.}$$

We next note that for each fixed $A \in \mathbf{A}$, the class of all subsets of X, $N(A \mid .)$ is a function defined on the population state space, while for fixed $x^n \in \Phi$, $N(. \mid x^n)$ is a function defined on $\mathbf{A}$ having the following properties:
  (i) it is non-negative and finite and integral-valued;
  (ii) it is completely additive in the sense that if $\{A_t, t \in T\}$ is any arbitrary collection of mutually disjoint members of $\mathbf{A}$, then there can be at most a finite number of sets $A_{t_1}, A_{t_2}, \ldots, A_{t_n}$
such that

$$N_{A_t} \geqslant 1$$

$$3.1.3 \qquad N\left(\sum_n A_t\right) = \sum_{i=1}^{n} N(A_{t_i}).$$

Thus we have a counting measure $\mathcal{N}$ on the set $\mathbf{A}$ of all subsets of X. In other words, the process $(\Phi, \mathbf{B}, P)$ generates a probability space $(\mathcal{N}, \mathbf{B}_n, P_N)$ which we shall call a *counting process*. It can be proved that there is a one-to-one correspondence between Φ and $\mathcal{N}$. The σ-field $\mathbf{B}_N$ can be identified to be the smallest σ-field of all sets $\{N \mid N(A_i) = k_i, i = 1, 2, \ldots, n\}$ in $\mathcal{N}$ where A_i is measurable and k_i is a non-negative integer.

To generate the probabilities P_N, we note that if $A_1, A_2, \ldots, A_l$ is a finite measurable partition of X and $k_1, k_2, \ldots, k_l$ are non-negative integers with $k_1 + k_2 + \cdots + k_l = n$, we have

$$P_N\{N(A_i) = k_i, i = 1, 2, \ldots, l\}$$

$$3.1.4 \qquad = \frac{n!}{k_1! \, k_1! \cdots k_l!} \, p^{(n)}(A_1^{k_1} \times A_2^{k_2} \times \cdots \times A_l^{k_l}),$$

where the combinatorial factor is due to the indistinguishable nature of the members of the population, and the probability distribution itself becomes symmetric.

A special case is of considerable interest and corresponds to the class of *compound processes* where, conditional upon the population size, the states of the individuals are independently distributed with the same distribution Q on $\mathbf{B}$; thus $P^{(n)}$ in this case is given by

$$3.1.5 \qquad p^{(n)} = p_n Q^{*n},$$

where p_n is the probability that the population is of size n and Q^{*n} is the nth product measure on $\mathbf{B}^n$ generated by Q. If we take p_n to be a Poisson distribution with parameter λ and take x to be an interval $[0, L]$ and the partitions also as disjoint intervals generated by the points $x_1, x_2, \ldots, x_l = L$, then 3.1.4 reduces to

$$P_N\{N(A_i) = k_i, i = 1, 2, \ldots, l\}$$

$$3.1.6 \qquad = \sum_{i=1}^{l} \left[\exp - \lambda(x_i - x_{i-1})\right]\left[\lambda(x_i - x_{i-1})^{k_i}/k_i!\right]$$

showing that the $N(A_i)$ are independent Poisson variables with expected value $\lambda(x_i - x_{i-1})$, $i = 1, 2, \ldots$.

3.2 Moment measures: product densities

Next we consider the *measures* generated by the moments of the counting process. Defining $M_1(A)$ for any $A \subset X$ by

$$3.2.1 \qquad M_1(A) = E[N(A)],$$

we notice that

$$3.2.2 \qquad M_1(A) = \sum_{n=1}^{\infty} \int_{x^n} N(A \mid x^n) P^{(n)}(dx^n).$$

Using the representation 3.1.1 and inverting the order of the summation and integration

$$3.2.3 \qquad M_1(A) = \int_{X^n} \sum_{i=1}^{n} N(A \mid x_i) P^{(n)}(dx^n).$$

Taking the measure $P^{(n)}(.)$ to be absolutely continuous and using the symmetric nature of the distribution, we obtain

$$M_1(A) = \int_A dx_1 \sum_{n=1}^{\infty} \frac{1}{(n-1)!} \int_{X^{n-1}} J_n(x_1, x_2, \ldots, x_n) dx_2 \cdots dx_n$$

$$\text{3.2.4} \qquad = \int_A dx_1 h_1(x_1),$$

where J_n is the probability density corresponding to the distribution $P^{(n)}(\,.\,)$ and $h_1(\,.\,)$ is easily identified:

$$\text{3.2.5} \qquad h_1(x_1) = \frac{1}{(n-1)!} \int_{X^{n-1}} J_n(x_1, x_2, \ldots, x_n) dx_2 \cdots dx_n.$$

The function $h_1(\,.\,)$ defined over X is known as the *product density of degree one* and is the derivative of the mean distribution of the measure N on $\mathbf{B}$.

Next we generalize 3.2.4 corresponding to higher moments of $N(A)$. At the outset we note that the moment distributions M_k $(k > 1)$ are not absolutely continuous and have mass concentrations on subsets of X^k. For instance, the second moment $M_2(A^{(2)})$, where $A^{(2)} \in \mathbf{B}^2$ is given by

$$\text{3.2.6} \qquad M_2(A^{(2)}) = M_{(2)}(A^{(2)}) + M(DA^{(2)}),$$

where $DA^{(2)}$ is the diagonal set consisting of $\{x : x \in X, x = y, (x, y) \in A^{(2)}\}$ and $M_{(2)}$ is given by

$$\text{3.2.7} \qquad M_{(2)}(A^{(2)}) = E[N_{(2)}(A^{(2)})]$$

$$\text{3.2.8} \qquad N_{(2)}(A^{(2)} | x^{(n)}) = \sum_{i_1 \neq i_2} \delta(A^{(2)} | x_{i_1}, x_{i_2}) \quad (n \geqslant 2).$$

Thus $N_{(2)}(\,.\, | x^{(n)})$ assigns measure one to every singleton element (x_{i_1}, x_{i_2}) such that $x_{i_1} \neq x_{i_2}, x_{i_1}, x_{i_2} \in \{x_1, x_2, \ldots, x_n\}$ and measure 0 to the rest of X^2. Hence we obtain

$$M_{(2)}(A^{(2)}) = \sum_{n=2}^{\infty} \int_{N^n} \sum_{i_1 \neq i_2} \delta(A^{(2)} | x_{i_1}, x_{i_2}) P^{(n)}(dx^n)$$

$$\text{3.2.9} \qquad = \int_{A^{(2)}} h_2(x_1, x_2) dx_1 dx_2$$

where

$$\text{3.2.10} \qquad h_2(x_1, x_2) = \sum_{n=2}^{\infty} \frac{n!}{(n-2)!} \int_{X^{n-2}} J_n(x_1, x_2, \ldots, x_n) dx_3 \cdots dx_n.$$

The quantity $h_2(x_1, x_2)dx_1 dx_2$ represents the expected value of the product measure $M_{(2)}(dx_1 \times dx_2)$. The function $h_2(\cdots)$ is called the *product density of degree* 2 or *second-order product density* of the population point process.

The formula 3.2.10 can be easily generalized. An explicit formula for $M_k(A^k)$ can be obtained by observing that $N_k(A^k | .)$ is related to $N_{(k)}(A^{(k)} | .)$ by

$$3.2.11 \qquad N_k(A^k | .) = \sum_l C_l^k N_{(l)}(A^{(l)} | .),$$

where C_l^k are the Stirling numbers of the second kind representing the $(k - l)$-fold degeneracy in a product of k terms due to coalescence. Hence we find

$$3.2.12 \qquad M_k(A^k) = \sum_l C_l^k M_{(l)}(A^{(l)})$$

where $M_{(l)}(A^{(l)})$ is now given by

$$3.2.13 \qquad M_{(l)}(A^{(l)}) = \int_{A^{(l)}} h_l(x_1, x_2, \ldots, x_l) dx_1 dx_2 \cdots dx_l,$$

where

$$h_l(x_1, x_2, \ldots, x_l) = \sum_{n=l}^{\infty} \left(\frac{n!}{(n-l)!} \right)$$

$$3.2.14 \qquad\qquad\qquad \times \int_{X^{n-l}} J_n(x_1, x_2, \ldots, x_n) dx_{l+1} \cdots dx_n.$$

The density function h_l is called the product density of degree l.

All along we have considered populations whose total size is finite with probability one; however, the theory leading to the explicit formulae for moments and product densities can be easily extended to cover the case when the population size is σ-finite. This leads to a σ-finite counting process. Thus the formulae given above are valid in the case when there can only be a finite number of members of the population with values in any bounded region of X. We now illustrate the main results of the theory by an example.

Example 3.2.1   Homogeneous birth and death process   We consider the following model defined over the time-interval $(0, T)$ $(T > 0)$. Initially at the time origin there is a cell. Each cell proliferates by binary fission at a constant rate λ per unit time, independently of other cells present. Likewise each cell dies at a constant rate μ per unit time, independently of other cells.

If we use the terminology population in a general sense, then the population of births over the time-interval $[0, T]$ is a population point process which we call a B-process. Likewise the population of deaths is a population point process, which we call a D-process. Now we proceed to obtain the product densities $h_1^B(t)$ and $h_1^D(t)$. These are in fact conditional densities conditional upon the presence of a cell at the time-origin.

To obtain an equation for $h_1^B(t)$ $(t > 0)$ we note that the cell present initially is subject to the following random course of events. The cell may, in the infinitesimal interval $(0, \Delta)$,

(i) undergo fission with probability $\lambda\Delta + o(\Delta)$ in which case we have two primary cells at the time-epoch Δ;

(ii) die with probability $\mu\Delta + o(\Delta)$ in which case the process becomes extinct.

There is a residual probability $1 - (\lambda + \mu)\Delta + o(\Delta)$ of the cells surviving the interval $(0, \Delta)$ without undergoing fission. Using the independent nature of the cells in the case of fission and the time-homogeneous nature of the fission and death probabilities, we obtain

$$h_1^B(t) = (1 - (\lambda + \mu)\Delta)h_1^B(t - \Delta)$$

3.2.15
$$+ 2\lambda\Delta h_1^B(t - \Delta) + o(\Delta)$$

which on taking the limit as $\Delta \to 0$ yields

3.2.16
$$\frac{d}{dt} h_1^B(t) = -(\mu - \lambda)h_1^B(t).$$

Observing that the initial condition is given by

3.2.17
$$h_1^B(0) = \lambda,$$

we obtain

3.2.18
$$h_1^B(t) = \lambda e^{-(\mu - \lambda)t}.$$

In an exactly similar way, we obtain

3.2.19
$$h_1^D(t) = \mu e^{-(\mu - \lambda)t}.$$

If we assume $\mu > \lambda$ and take $T = +\infty$, we have the birth- and death-point process on the positive real line. This corresponds to the situation when the population size is finite with probability 1. If $M^B(R^+)$ is the expected size of the population of the B-process,

3.2.20
$$M^B(R^+) = \lambda/(\mu - \lambda).$$

Likewise we have for the D-process

3.2.21 $\quad M^D(R^+) = \mu/(\mu - \lambda).$

If $\mu < \lambda$, then we have a σ-finite population.

Higher-order product densities can be obtained in a similar way. For instance, the function $h_2^{BB}(t_1, t_2)$ satisfies for arbitrarily small $\Delta > 0$

$$h_2^{BB}(t_1, t_2) = \{1 - (\lambda + \mu)\Delta\}h_2^{BB}(t_1 - \Delta, t_2 - \Delta)$$
$$+ 2\lambda\Delta h_2^{BB}(t_1 - \Delta, t_2 - \Delta)$$

3.2.22 $\qquad\qquad + 2\lambda\Delta h_1^B(t_1 - \Delta)h_1^B(t_2 - \Delta) + o(\Delta)$

which on taking the limit as $\Delta \to 0$ yields

$$\left(\frac{\partial}{\partial t_1} + \frac{\partial}{\partial t_2}\right)h_2^{BB}(t_1, t_2) + (\mu - \lambda)h_2^{BB}(t_1, t_2)$$

3.2.23 $\qquad = 2\lambda h_1^B(t_1)h_1^B(t_2).$

The above equation can be solved by observing that in the present case the boundary condition is given by

3.2.24 $\quad h_2^{BB}(0, t_2) = 2\lambda h_1^B(t_2) \quad (t_2 > 0).$

Equations for other types of product densities can be written down and solved. However, we can obtain explicit expressions for these densities by the use of combinatorial arguments as follows. We first deal with h_2^{BB}. Let $N(t_1)$ represent the size of the population of cells at epoch t_1. Then it is easy to see the conditional value of $h_2^{BB}(t_1, t_2)$ conditional on the number of cells at epoch t_1 for $t_1 < t_2$ is given by

$$h_2^{BB}(t_1, t_2) \big| \text{conditional}$$

3.2.25 $\qquad = \lambda N(t_1)[N(t_1) + 1]h_1^B(t_2 - t_1).$

The above equation follows if we observe that with a birth in an infinitesimal interval following t_1, the population size of cells is $N(t_1) + 1$ and each of the cells can cause a birth in an infinitesimal interval following t_2. Thus we finally have

3.2.26 $\quad h_2^{BB}(t_1, t_2) = \lambda h_1^B(t_2 - t_1)E[N(t_1)[N(t_1) + 1]].$

The expected value of the quantity within the bracket on the r.h.s. of 3.2.26 can be expressed in terms of the second factorial moment and the first moment of $N(t_1)$, expressions for which are already available (see Example

2.2.5):

3.2.27 $$E[N(t_1)] = e^{-(\mu-\lambda)t_1}$$

$$E\{N(t_1)[N(t_1)-1]\}$$

3.2.28 $$= \frac{2\lambda}{(\mu-\lambda)} e^{-(\mu-\lambda)t_1}(1 - e^{-(\mu-\lambda)t_1}).$$

Hence we have

3.2.29 $$h_2^{BB}(t_1, t_2) = \frac{2\lambda^2}{\mu-\lambda} e^{-(\mu-\lambda)t_2}[\mu - \lambda e^{-(\mu-\lambda)t_1}].$$

It is easily verified that the r.h.s. satisfies the differential equation 3.2.22 and the boundary condition 3.2.24.

Using the same type of argument, it is easy to obtain explicit expressions for the product density of the D-process:

$$h_2^{DD}(t_1, t_2) = \mu h_1^D(t_2 - t_1)E\{N(t_1)[N(t_1)-1]\}$$

3.2.30 $$= \frac{2\lambda\mu^2}{\mu-\lambda} e^{-(\mu-\lambda)t_2}(1 - e^{-(\mu-\lambda)t_1}).$$

Higher-order densities can be obtained in a similar way and perhaps one can arrive at a recursive relation.

In the above example we dealt with two types of processes from a general setting; this can be viewed as a population point process with vector type of entities. Such a point process is known under various names: *multivariate point process, marked point process* or *multi-type series of events*. For instance, the births and deaths in the above example are two distinct types of events, and the marginal processes of B and D events are not independent. The main results presented above are equally valid in such a context. In this case we encounter *multivariate product densities* which arise from the measures corresponding to the counting process of events of either type.

To ensure clarity we pursue Example 3.2.1 a little further and note that $h_2^{DB}(t_1, t_2)$ and $h_2^{BD}(t_1, t_2)$ are the mixed product densities. Explicit expressions for them follow by combinatorial arguments used earlier. We thus find for $t_1 > t_2$

$$h_2^{BD}(t_1, t_2) = E\{N(t_1)[N(t_1)+1]\}h_1^D(t_2 - t_1)$$

3.2.31 $$= \frac{2\lambda\mu}{(\mu-\lambda)} e^{-(\mu-\lambda)t_2}[\mu - \lambda e^{-(\mu-\lambda)t_1}],$$

$$h_2^{DB}(t_1, t_2) = \mu E\{N(t_1)[N(t_1) - 1]\}h_1^{B}(t_2 - t_1)$$

3.2.32
$$= \frac{2\lambda^2\mu}{(\mu - \lambda)} e^{-(\mu - \lambda)t_2}[1 - e^{-(\mu - \lambda)t_1}].$$

We note that the functions h_1^{B}, h_1^{D}, h_2^{DD}, h_2^{DB} and h_2^{BB} provide a second-order description of the point process in question. If we confine our attention to the finite interval $I = [0, T]$, the first two factorial moments of the counting measure are given by

3.2.33
$$M^{B}(I) = \frac{\lambda}{\mu - \lambda}\{1 - e^{-(\mu - \lambda)T}\}.$$

3.2.34
$$M^{D}(I) = \frac{\mu}{\mu - \lambda}\{1 - e^{-(\mu - \lambda)T}\}.$$

3.2.35
$$M_{(2)}^{B}(I) = \frac{4\lambda^2}{(\mu - \lambda)^2}\left\{1 - e^{-(\mu - \lambda)T} - \mu T e^{-(\mu - \lambda)T} \right. $$
$$\left. + \frac{\lambda}{2(\mu - \lambda)}[1 - e^{-2(\mu - \lambda)T}]\right\}.$$

3.2.26
$$M_{(2)}^{D}(I) = \frac{4\lambda^2\mu}{(\mu - \lambda)^2}\left\{\frac{1 - e^{-2(\mu - \lambda)T}}{2(\mu - \lambda)} - T e^{-(\mu - \lambda)T}\right\}.$$

The cross-correlation measure M^{BD} is given by

3.2.37
$$M^{BD}(I) = \frac{\mu}{2\lambda} M_{(2)}^{B}(I) + \frac{\lambda}{2\mu} M_{(2)}^{D}(I).$$

Higher-order multivariate product densities can be defined, and explicit expressions for the higher moments of measures obtained.

Example 3.2.2   Electron–photon cascade   In cosmic-ray physics, the following model of multiplicative process is studied. A highly energetic photon is incident on the top layer of the atmosphere. Due to interaction with matter (very often artificially introduced in the form of an experiment in which a thick sandwich of emulsion is presented to the incoming gamma ray/other particles), the photon gives rise to an electron–positron pair. The electron or the positron in turn interacts with matter present and emits a photon by the process of bremsstrahlung. The process can be represented as follows:

$$\gamma \to e^{+} + e^{-}, \quad e^{\pm} \to e^{\pm} + \gamma,$$

where γ stands for photon and e^{+} and e^{-} represent electrons and positrons. In a certain approximation, the quantum mechanical cross-section for these processes can be obtained; the probability per unit thickness that an electron (or a positron) of energy E gives rise to a photon with energy lying in the interval $(E', E' + dE')$ is given by $R_1(E, E')dE'$, where

$$3.2.38 \qquad R_1(E, E') = \frac{\frac{4}{3}(E^2 - EE') + E'^2}{\left[1 + \omega^2\left(\dfrac{E}{E'}\right)^2\right]E^2 E'} \qquad (E' \leqslant E),$$

where ω is a factor depending on the polarization characteristics of the material. The values of ω are 1.9×10^{-4} and 7.5×10^{-5} for air and lead respectively. The probability per unit thickness (of matter) that a photon of energy E gives rise to an electron–positron pair one of which has an energy lying in $(E', E' + dE')$ is given by $R_2(E, E')dE'$, where

$$3.2.39 \qquad R_2(E, E') = \left\{1 - \frac{4}{3}\left(\frac{E'}{E} - \frac{E'^2}{E^2}\right)\right\}\frac{1}{E} \qquad (E' \leqslant E).$$

For our purposes, electrons and positrons behave in the same way and so we shall simply use the term electron to denote either. If an electron (a photon) of energy E_0 is incident on the top layer, it produces further photons (annihilates into an electron pair) which in turn produce further particles in the passage through matter, resulting in a *shower* of particles. The amount of thickness penetrated is the only parameter, and we shall treat the problem as one-dimensional; if we use t to denote thickness, it is easy to see that we have a population of electrons and photons of various energies $(< E_0)$ at any $t > 0$. In other words, for any t and given energy E_0 we have a population point process over the real interval $[0, E_0]$. In view of the assumptions regarding the cross-sections, there is a tendency for the particles to multiply, with the energies decreasing progressively. There are several interesting stochastic point processes; the population at thickness $t > 0$ of electrons and photons generated by an electron or a photon of energy E_0 initially (at $t = 0$) is a bivariate population point process over the real interval $[0, E_0]$. The points (layers) at which the particles are produced form an interesting point process; thus we have a population point process over the product space $[0, E_0] \times [0, T]$, where T is any arbitrary positive number. There is yet another interesting point process; this arises when we take into account the state of polarization of electrons and photons. Since the particles are highly energetic, the initial particles are expected to be completely polarized; thus we can reasonably take the particles to be polarized forward (F) or backward (B) relative to the direction of motion. Of course the quantum mechanical cross-sections R_1 and R_2 which are averaged over the spin states have to be defined in a finer manner. Then it can be shown that there is an interesting phenomenon of polarization cascades. Thus the population of particles at a particular thickness t is a point process over the product space $\{F, B\} \times [0, E_0]$.

We first consider the point process over $[0, E_0]$ for a fixed $t > 0$. The first-order product densities can be denoted by $h_1^i(j, E, t \mid E_0)$, where the superscript i refers to the primary (present initially) generating the cascade and takes the value 1 or 2 according as the shower is excited by an electron or a photon respectively. The values $j = 1, 2$ correspond respectively to the product densities of electrons and photons. Next we observe that in the infinitesimal layer $(0, \Delta)$ the incident primary of type i may collide with one of the atoms of the material layer to produce a secondary with energy in $(E', E' + dE')$ with probability $R_i(E, E')dE'\Delta$, the probability of no collision in $(0, \Delta)$ being residual. Hence we obtain for arbitrarily small $\Delta > 0$

$$h_1^i(j, E, t \mid E_0)$$

$$= \Delta \int_0^{E_0} R_i(E_0, E')[h_1^{3-i}(j, E, t - \Delta \mid E')$$

$$+ h_1^i(j, E, t - \Delta \mid E_0 - E')]dE'$$

$$3.2.40 \qquad + h_1^i(j, E, t - \Delta \mid E_0)\left\{1 - \Delta \int_0^{E_0} R_i(E_0, E')dE'\right\} + o(\Delta).$$

Proceeding to the limit as $\Delta \to 0$, we obtain

$$\frac{\partial}{\partial t} h_1^i(j, E, t \mid E_0)$$

$$= \int_0^{E_0} R_i(E_0, E')[h_1^{3-i}(j, E, t \mid E')$$

$$+ h_1^i(j, E, t \mid E_0 - E')]dE'$$

$$3.2.41 \qquad - h_1^i(j, E, t \mid E_0) \int_0^{E_0} R_i(E_0, E')dE',$$

the equation being valid for all values of i and j. The initial conditions are given by

$$3.2.42 \qquad h_1^i(j, E, t \mid E_0) = \delta(E - E_0)\delta_{ij},$$

where $\delta(.)$ is the Dirac delta function. From the form of the functions

$R_i(E, E')$ specified by 3.2.38 and 3.2.39 we can infer that the set of equations 3.2.40 and 3.2.41 are amenable to Mellin transform technique. The steps are very similar to those outlined in Example 2.2.6.

If we consider the population point process of production of particles and denote the product densities by $h_1^i(j, E, t \mid E_0)$ by a misuse of the notation, then it is easy to see that equation 3.2.41 still holds; however, the initial conditions 3.2.42 have to be replaced by

3.2.43 $\qquad h_1^1(j, E, t \mid E_0) = R_1(E_0, E)\delta_{j2} + R_1(E_0, E_0 - E)\delta_{j1},$

3.2.44 $\qquad h_1^2(j, E, t \mid E_0) = 2R_2(E_0, E)\delta_{j1}.$

Equations for the multi-product densities of second order can be set up in a similar way. It is indeed possible to estimate the second moment of the population size. (See Srinivasan (1969), chapters 5 and 6.)

In the two examples presented, it is worth noting that the product densities of the population point process satisfy the Chapman–Kolmogorov relation. The validity of the relation can be established in different ways. The product densities are nothing but expected values, and the physical situation demands a conservation (continuity) equation to be satisfied. Yet another way of establishing the validity stems from the interpretation of product density as the derivative of probability measure implied by 3.2.1. This has been widely recognized from the very beginning (see for example Bartlett, 1955). A formal and more rigorous way of establishing such equations consists in recognizing the Markov nature of the process which results in the Chapman–Kolmogorov equation satisfied by the probability measure $P^{(m)}(.)$ or by the corresponding derivative $J_n(x_1, x_2, \ldots, x_n)$; at that stage if we use 3.2.14, we are then led to the Chapman–Kolmogorov equation for the product densities.

3.3 Stationary point processes

So far we have studied the point process as a non-negative integer-valued measure on the σ-algebra of subsets of the reals. We have seen that it gives rise to point events on the real line and hence we can visualize the process as the collection of intervals between the successive events. For instance, the function $J_n(x_1, x_2, \ldots, x_n)$ characterizes the joint distribution of the intervals, if X happens to be a bounded interval. On the other hand, if we take X to be the whole real line, we can visualize the point process as a sequence of the location of point events

$$\{t_n : n = \pm 1, \pm 2, \ldots : \ldots t_{-2} < t_{-1} < 0 = t_0 < t_1 < t_2 \cdots\}.$$

The original assignment of measures to subsets implies that $X_i = t_i - t_{i-1}$, $i \in I$ is a stochastic process. The values corresponding to 1 and 0 do not correspond to intervals; nevertheless, as we shall see below, they play an important role in the development of the theory. In what follows we call the sequence $\{t_n\}$ the point process.

Example 3.3.1   Poisson process   Let $\{X_i, i \in I, i \neq 0, 1\}$ be a collection of independent and exponentially distributed random variables with parameter $\lambda > 0$. If we define $N(t)$ for $t > 0$ by

3.3.1 $N(t) = \sup\{n : X_1 + X_2 + \cdots + X_n \leqslant t\}$

then the counting process $\{N(t), t > 0\}$ reduces to Example 2.1.3 and hence represents the simple Poisson process. The epochs $\{t_i : i = \pm 1, \pm 2, \ldots\}$ represent the point events, t_i being measured w.r.t. an arbitrary origin $t_0 = 0$. It is to be specially noted that the exponential nature of the distribution of the intervals between successive events leads to the statistical independence of X_1 and X_{-1}; the variables X_1 and X_{-1} are also exponentially distributed with the same parameter λ.

Example 3.3.2   Renewal process   In Example 3.3.1 we removed the assumption of exponential distribution. The resulting process is called a *renewal process*. The random variable X_1 is called the *forward recurrence time*; it represents the time to the first event measured from an arbitrary origin. Likewise the random variable X_{-1} is called the *backward recurrence time*.

Stationarity

The process $\{t_n\}$ is called a *completely (strictly) stationary process* if the counting process $N(t)$ representing the counting measure of the interval $(0, t)$ is completely (strictly) stationary. There are of course weaker versions of stationarity but we will not have occasion to deal with them. The Poisson process described in Examples 2.1.3 or 3.3.1 is completely stationary. The renewal process as described by Example 3.3.2 can be made completely stationary provided we define the dependency structure of X_1 and X_{-1}. For a detailed account of renewal processes, we refer the reader to Cox (1962).

Next we note that stationarity implies that $E[N(t)]$ is a linear function of

t so that we have

3.3.2 $\quad E[N(t)] = \mu t,$

where μ is a positive constant characteristic of the process in question. In the case of the simple Poisson process, μ is the parameter of the exponential distribution governing the intervals X_i. It is worth noting that

3.3.3 $\quad E[X_i] = 1/\mu.$

Next we study the implication of stationarity on the product densities. We note that the first-order product density is the density function of the expected counting measure evaluated at x_1, and hence it follows from 3.2.1 that

3.3.4 $\quad h_1(x_1) = \lim_{\Delta \to 0} E[N(A)]/\Delta$

where A is the interval $(x_1, x_1 + \Delta)$. Thus stationarity implies that the value $h_1(x_1)$ must be independent of x_1 so that we have

3.3.5 $\quad h_1(x_1) = $ a constant $= \mu.$

On the other hand, stationarity also implies that

3.3.6 $\quad Pr\{N(A) \geqslant 1\} = \sum_{i \geqslant 1} Pr\{N(A) = i\} = \lambda\Delta + o(\Delta).$

If we assume that multiple events or clusters of events do not occur with probability one, then we have

3.3.7 $\quad E[N(A)] = Pr\{N(A) = 1\} + o(\Delta)$

so that $\lambda = \mu$. A process in which multiple events do not occur with probability one is called a *regular process* or an *orderly process* since in such a case the events can be ordered in a sequence. Thus we have the result *that a stationary point process is regular if and only if* $\mu = \lambda$; this result is known as *Korolyuk's theorem*. In what follows we assume that the process is regular.

Stationarity also implies that the second-order product density $h_2(x_1, x_2)$ for $x_1 < x_2$ is a function of $x_2 - x_1$ only. In this case we can write

3.3.8 $\quad h_2(x_1, x_2) = \mu h_1^c(x_2 - x_1),$

where $h_1^c(.)$ is the conditional produce density arising from conditional expected measure and is defined by

3.3.9 $\quad h_1^c(x) = \lim_{\Delta, \Delta' \to 0} E[N(A)\,|\,N(B) = 1]/\Delta,$

where A and B are respectively the intervals $(x, x + \Delta)$ and $(-\Delta', 0)$. These results simplify the moment formula 3.2.12 considerably. We have, if A is taken to be the interval $[a, b]$,

$$3.3.10 \qquad E[M_2(A^{(2)})] = \mu(b - a) + 2\mu \int_0^{b-a} h_1^c(x)(b - a - x)\,\mathrm{d}x.$$

In our discussion so far we have made no mention of the choice of the origin. There are two possibilities for a stationary point process. The process can be started at $t = 0$ with appropriate initial conditions that produce stationarity. Such initial conditions are called *stationary initial conditions*. Let us amplify the point a little further. In the case of the renewal process introduced in Example 3.3.2, X_1 is distributed differently from X_i $(i > 1)$. If we choose the density function $f_1(.)$ of X_1 as

$$3.3.11 \qquad f_1(x) = [1 - F(x)]/\mu,$$

where μ is the expected value of X_i $(i > 1)$ and $F(.)$ the distribution function of X_i $(i > 1)$, then the process $\{N(t); t > 0\}$ becomes a stationary renewal process. The alternative way is to take the process to be transient and visualize it beyond $t_0 = 0$ as the starting-point moves off to the left. As the starting-point recedes to minus infinity, the process becomes stationary. The specification of the process at $t = 0$ is known as the *stationary equilibrium condition*. We illustrate this by an example which will be quite useful later on.

Example 3.3.3 The point process of births/deaths Let us consider Example 3.2.1; the point process of births/deaths is not stationary. To obtain a stationary process, we impose on the process an immigration process with a constant (expected) rate of immigration v. In other words, we have a Poisson process of immigration with parameter v and each member of the population generates independent of other members a birth-and-death process as in Example 3.2.1. We adopt the second method mentioned above by receding the starting-point to minus infinity. To achieve this, we just note that the product densities of birth/death processes obtained in Example 3.2.1 are immigration taboo product densities conditional on one cell being present at the origin. For notational convenience we can pre-subscript them by the symbol I; thus we have the product densities $_Ih_1^B(.)$, $_Ih_1^D(.)$, $_Ih_2^{BB}(.)$, $_Ih_2^{DD}(., .)$, $_Ih_2^{BD}(.)$ and $_Ih_2^{DB}(., .)$, specified respectively by 3.2.18, 3.2.19, 3.2.29, 3.2.30, 3.2.31 and 3.2.32. Suppose there are no cells at the starting-point $(t = 0)$; the product densities corresponding to this initial condition, and with immigration no longer taboo, will be denoted by the same symbols with the pre-subscript I deleted. We now consider the time-interval $(0, \Delta)$ for arbitrarily small $\Delta > 0$; there are two possibilities:

(i) a cell enters the system by immigration with probability $v\Delta + o(\Delta)$;

(ii) the initial conditioning continues to remain the same with the residual probability $1 - v\Delta + o(\Delta)$. In the case of (i), the contribution to the first-order product density arising from a single cell present at Δ is $_I h_1^B(t - \Delta)$; there is also an additional contribution due to immigration over the interval (Δ, t) and this is simply $h_1^B(t - \Delta)$. We thus have:

$$3.3.12 \qquad h_1^B(t) = v\Delta\{_I h_1^B(t - \Delta) + h_1^B(t - \Delta)\} + (1 - v\Delta)h_1^B(t - \Delta) + o(\Delta)$$

which leads to

$$3.3.13 \qquad \frac{\mathrm{d}}{\mathrm{d}t} h_1^B(t) = v_I h_1^B(t).$$

In a similar manner we have

$$3.3.14 \qquad \frac{\mathrm{d}}{\mathrm{d}t} h_1^D(t) = v_I h_1^D(t).$$

The initial conditions are

$$3.3.15 \qquad h_1^D(0) = h_1^B(0) = 0.$$

Using the formula 3.2.18 and 3.2.19 we obtain

$$3.3.16 \qquad h_1^B(t) = \frac{v\lambda}{\mu - \lambda}\left[1 - \mathrm{e}^{-(\mu - \lambda)t}\right]$$

$$3.3.17 \qquad h_1^D(t) = \frac{v\mu}{\mu - \lambda}\left[1 - \mathrm{e}^{-(\mu - \lambda)t}\right].$$

To obtain the second-order product densities, we proceed in the same way. Thus we have

$$h_2^{BB}(t_1, t_2) = \{_I h_2^{BB}(t_1, t_2) + h_2^{BB}(t_1, t_2)\}v\Delta$$

$$3.3.18 \qquad\qquad\qquad + (1 - v\Delta)h_2^{BB}(t_1 - \Delta, t_2 - \Delta) + o(\Delta)$$

which leads to

$$3.3.19 \qquad \left(\frac{\partial}{\partial t_1} + \frac{\partial}{\partial t_2}\right)h_2^B(t_1, t_2) = v_I h_2^{BB}(t_1, t_2).$$

The initial condition can be taken to be

3.3.20 $\qquad h_2^B(0, t_2) = 0.$

Solving the above partial differential equation, we obtain

$$h_2^{BB}(t_1, t_2) = \frac{v\lambda^2}{(\mu - \lambda)^2} \, e^{-(\mu - \lambda)(t_2 - t_1)}$$

3.3.21 $\qquad \times \{(2\mu - \lambda) - 2\mu\, e^{-(\mu - \lambda)t_1} + \lambda\, e^{-2(\mu - \lambda)t_1}\}.$

We next observe that receding the time-origin to $-\infty$ is equivalent to taking the limit as $t_1 \to \infty, t_2 \to \infty$ keeping $t_2 - t_1$ fixed. Under this limit we have

3.3.22 $\qquad h_1^B(\,.\,) = \dfrac{v\lambda}{\mu - \lambda}$

3.3.23 $\qquad h_1^D(\,.\,) = \dfrac{\mu v}{\mu - \lambda}$

3.3.24 $\qquad h_2^{BB}(t_1, t_2) = \dfrac{v\lambda^2(2\mu - \lambda)}{(\mu - \lambda)^2} \, e^{-(\mu - \lambda)(t_2 - t_1)}.$

An exactly similar method yields explicit formulae for h_2^{DD}, h_2^{BD} and h_2^{DB}. This procedure establishes the stationarity of the process at least up to second order.

Finally, returning to the sequence $\{t_n\}$ or the collection $\{X_i\}$, we note that stationarity has a direct consequence on the distribution of X_1 and X_{-1}. To obtain the distribution of X_1, we can proceed as follows, using rather direct intuitive arguments. We assume stationary equilibrium condition at the time-origin; we trace the time backwards until the time-point t_{-1} when a count is realized for the counting process. Observing that the time-interval between t_{-1} and t_1 is governed by the stationary distribution of the interval between two successive events, we find

$$Pr\{x < X_1 < x + dx\} = \lambda \int_0^\infty f(x + y)\,dy\,dx$$

3.3.25 $\qquad = \lambda(1 - F(x))dx,$

where $F(\,.\,)$ is the stationary distribution of the interval. The rate of the process equals the reciprocal of the first moment of the stationary interval distribution:

$$3.3.26 \qquad 1/\lambda = \int_0^\infty [1 - F(x)]\,dx = \mu_1.$$

In fact, the same type of argument can be used to arrive at the joint probability density function of X_{-1} and X_1. If $g_{-1,1}(x, y)$ is the joint density function, it is easy to prove that

$$3.3.27 \qquad g_{-1,1}(x, y) = \lambda f(x + y),$$

from which we conclude that X_1 and X_{-1} are identically distributed. We can prove more generally that the nth forward and backward recurrence times are identically distributed. It is worth noting that in the case of a Poisson process, the common distribution of X_1 and X_{-1} coincides with $F(.)$. We recall that a Poisson process is a process of stationary independent increments. If we relax the constraint of independence we obtain a stationary point process.

The determination of the structure of the distribution of the nth forward/backward recurrence time is a very difficult problem and is generally done by discovering some embedded renewal process. In the case of renewal processes the distribution of the nth forward/backward recurrence-time takes a simple form in view of the independent nature of the intervals. This, if we denote by $f_n(.)$ the p.d.f. of the nth recurrence time, we have

$$3.3.28 \qquad f_n(x) = \lambda \bar{F} * f^{(n-1)}(x),$$

where $f^{(n-1)}$ is the $(n-1)$fold convolute of f and $\bar{F} * g^{(n-1)}$ is the convolute of $\bar{F}$ with $f^{(n-1)}$. In terms of Laplace transforms we have

$$3.3.29 \qquad f_n^*(s) = \lambda[f^*(s)]^{n-1}(1 - f^*(s))/s.$$

The independent nature of the intervals also introduces a simplicity in the structure of the product densities. In view of the independent nature of the random variables X_i $(i \neq -1)$, we have for $x_1 < x_2 < \cdots < x_n$

$$h_n(x_1, x_2, \ldots, x_n) = \mu h_1^c(x_2 - x_1)$$

$$3.3.30 \qquad\qquad\qquad \times h_1^c(x_3 - x_2) \cdots h_1^c(x_n - x_{n-1})$$

where $h_1^c(.)$ is the conditional product density defined by 3.3.9. A further use of the independent nature of the random variables gives

$$3.3.31 \qquad h_1^c(x) = f(x) + \int_0^x f(y) h_1^c(x - y)\,dy$$

which in turn reduces to

$$3.3.32 \qquad h_1^{c*}(s) = f^*(s)/[1 - f^*(s)].$$

The conditional product density $h_1^c(\,.\,)$ is known as the *renewal density* for the simple reason that it represents the expected intensity or density of the point events. A renewal process can be characterized in terms of just the renewal density since there is a one-to-one correspondence between the functions $f(\,.\,)$ and $h_1^c(\,.\,)$.

3.4 Cox processes

If we consider a stationary Poisson process the parameter characterizing the process has several interpretations:

(i) It is the parameter of the common exponential distribution of the collection of independent random variables $\{X_n\}$ that define the process.

(ii) It is the probability per unit time of the occurrence of events (which are independent):

$$3.4.1 \qquad \lambda = \lim_{\Delta \to 0} Pr\{N(A) = 1\}/\Delta$$

where A is the interval $(t, t + \Delta)$.

The Poisson character as well as the interpretation of the parameter 3.4.1 remain valid if λ is replaced by a positive definite function $\Lambda(t)$; however, in this case the process is non-stationary and hence we need to change the notation slightly. For instance, the probability distribution of the forward recurrence time measured from an arbitrary point t depends on t as well. The counting process which can now be denoted by $N([t, t + x])$ has many characteristics in common with the stationary Poisson process. For instance, the probability mass function of $N((t, t + x])$ denoted by $p(n, t, x)$ is now given by

$$3.4.2 \qquad p(n, t, x) = [\exp -\alpha(t, x)] \frac{[\alpha(t, x)]^n}{n!}$$

where

$$3.4.3 \qquad \alpha(t, x) = \int_t^{t+x} \Lambda(u)\,du.$$

The random variables X_1 and X_{-1} which depend on t are now no longer identically distributed. Denoting the density functions by $g_1(t, x)$ and $g_{-1}(t, x)$, we have

3.4.4 $\qquad g_1(t, x) = \Lambda(t + x)[\exp -\alpha(t, x)]$

3.4.5 $\qquad g_{-1}(t, x) = \Lambda(t - x)[\exp -\alpha(t - x, t)]$.

The product densities retain their general character; the product density of degree n is now given by

3.4.6 $\qquad h_n(x_1, x_2, \ldots, x_n) = h_1(x_1)h_1(x_2) \cdots h_1(x_n)$

3.4.7 $\qquad h_1(x) = \Lambda(x)$.

Thus all the properties of the simple Poisson process carry over except those that spell out independence of the time-origin.

If at this stage we take $\Lambda(t)$ to be the sample path of a stationary stochastic process $\Lambda(t, \omega)$, the probability measures introduced above can be interpreted as conditional probabilities, given $\{\Lambda(t, \omega) : t \in R\}$. We then realize a stationary point process known as the *doubly stochastic Poisson process* (DSPP) or *conditioned Poisson process* (CPP). The process is also known as the *Cox process*. Generally the process $\Lambda(t, \omega)$ is taken to be independent of the counting process; when it is dependent on the counting process as well, it is known as a *self-inhibited* or *self-exciting process*.

At the outset we observe that the very fact that the conditioned process is a Poisson process introduces an enormous amount of simplicity in the approach. The various probabilities or densities specified by 3.4.2 through 3.4.7 are now conditional probabilities. The many-time generating function of the counting process, corresponding to the DSPP for the disjoint intervals $[t_i, t_i + T_i]$, $i = 1, 2, \ldots, m$, is given by

$$\Phi_m(z_1, t_1, T_1; z_2, t_2, T_2; \cdots z_m, t_m, T_m)$$

3.4.8 $$= E_\Lambda\left[\exp \sum_{i=1}^{m} (z_i - 1) \int_{t_i}^{t_i + T_i} \Lambda(\tau) d(\tau)\right],$$

where the expectation is with respect to the process $\Lambda(.)$. The determination of the many-time generating function is a difficult problem in general unless the process $\Lambda(t)$ has amenable characteristics. For Cox processes it is a fairly easy job to derive explicit expressions for the product densities of the first few orders. In the case of *self-inhibited* or *self-exciting processes* we may have to use indirect methods to obtain the characteristics. We conclude the section by a few illustrative examples.

Example 3.4.1   A Cox process driven by the modulus square of a stationary Gaussian process   Let $\{X(t)\}$ be a stationary complex Gaussian process with

3.4.9 $E[X(t)] = 0$

3.4.10 $\text{cov}[X(t), X(s)] = R(t - s); R(0) = I > 0$

3.4.11 $\Lambda(t) = |X(t)|^2,$

where I is suggestive of the intensity of the resulting point process. An expression for the first-order product density follows quite easily. We have

3.4.12 $h_1(t) = E[|\Lambda(t)|^2] = I.$

The second-order product density is given by

3.4.13 $h_2(t_1, t_2) = E[\Lambda^+(t_1\Lambda(t_1)\Lambda^+(t_2)\Lambda(t_2)].$

The expression on the r.h.s. is easily evaluated using the Gaussian nature of the process; thus we have

3.4.14 $h_2(t_1, t_2) = I^2 + |R(t_1 - t_2)|^2.$

Higher-order product densities are obtained in a similar way. The generating function $\Phi_1(z, \tau, T)$ corresponding to the counting process over the interval $[\tau, \tau + T]$ is given by

$$\Phi_1(z, \tau, T) = E \exp(z - 1) \int_\tau^{\tau + T} |X(t)|^2 dt$$

3.4.15
$$= E\left[\exp(z - 1) \int_0^T |X(t)|^2 dt \right]$$

by virtue of stationarity of the process $\{X(t)\}$. We can use the Karuhnen–Loeve representation 2.1.19 in this case. Thus

3.4.16 $X(t) = \sum \phi_n(t)Z_n \quad (0 \leqslant t \leqslant T)$

where $\{Z_n\}$ is a family of uncorrelated Gaussian random variables with

3.4.17 $E[Z_n] = 0, \quad E[Z_n\bar{Z}_n] = \lambda_n,$

where the λ_n's are the eigen-values corresponding to the eigen-functions $\phi_n(t)$ of the integral equation

3.4.18 $\int_0^T R(t - s)\phi(s)ds = \lambda\phi(t).$

Using the fact that $\{\phi_n(t)\}$ is an orthonormal family over the interval $[0, t]$, we can evaluate the expectation on the r.h.s. of 3.4.15. Thus we obtain

3.4.19 $\phi_1(z, \tau, T) = \Phi_1(z, T) = \prod_k (1 + (1 - z)\lambda_k)^{-1}.$

Thus the problem reduces to one of determining the infinite product on the r.h.s. of 3.4.19.

Let us consider a special case when

3.4.20 $R(t - s) = I\,e^{-\Gamma|t - s| + i\omega(t - s)}$

which in turn corresponds to $\{X(t)\}$ being Markov. In Example 2.1.4 we have obtained the eigen-functions corresponding to the autocorrelation function defined by 3.4.20. We follow the same method. Introducing $f(s)$ by

3.4.21 $f(s) = \phi(s)\,e^{-i\omega s}$

the integral equation can be written as

3.4.22 $I \int_0^T e^{-\Gamma|u - v|} f(v)\,dv = \lambda f(u).$

We next introduce the Laplace transform $f^*(p)$ by

3.4.23 $f^*(p) = \int_0^T f(u)\,e^{-pu}\,du$

we obtain after some computations

3.4.24 $f^*(p) = I[(\Gamma - p)\,e^{-(\Gamma + p)T} f^*(-\Gamma)$

$$+ (\Gamma + p)f^*(\Gamma)]/[2\Gamma I - \lambda(\Gamma^2 - p^2)].$$

The denominator of the r.h.s. of 3.4.24 has two zeros, say p_1 and p_2. However, since $f^*(p)$ is analytic, the numerator of the r.h.s. of 3.4.24 vanishes when $p = p_1, p_2$. Thus we have

3.4.25 $(\Gamma - p_1)\,e^{-(\Gamma + p_1)T} f^*(-\Gamma) + (\Gamma + p_1)f^*(\Gamma) = 0$

3.4.26 $(\Gamma - p_2)\,e^{-(\Gamma + p_2)T} f^*(-\Gamma) + (\Gamma + p_2)f^*(\Gamma) = 0,$

which in turn implies that D the determinant of the coefficients of $f^*(-\Gamma)$ and $f^*(\Gamma)$ vanishes in order that $f^*(\Gamma)$ and $f^*(-\Gamma)$ may be non-trivial:

3.4.27 $D = \begin{vmatrix} (\Gamma - p_1)\,e^{-(\Gamma + p_1)T} & \Gamma + p_1 \\ (\Gamma - p_2)\,e^{-(\Gamma + p_2)T} & \Gamma + p_2 \end{vmatrix} = 0.$

The determinant D is a function of λ; however, for convenience we use the notation $D(\xi)$ where $\xi = 1/\lambda$; equation 3.4.27 is the eigen-value equation. The roots p_1 and p_2 are given by

3.4.28 $\qquad p_1 = -p_2 = (\Gamma^2 - 2\Gamma I/\lambda)^{1/2} = (\Gamma^2 - 2\xi I\Gamma)^{1/2}.$

Now the function $P(\xi)$ defined by

3.4.29 $\qquad P(\xi) = D(\xi)/(\Gamma^2 - 2\Gamma I\xi)^{1/2}$

is an entire function in the ξ-plane. The order of the entire function is easily seen to be half; its zeros are $1/\lambda_k$ by construction. By the Hadamard factorization theorem it follows that

3.4.30 $\qquad P(\xi) = P(0) \prod_k (1 - \xi/\xi_k).$

Hence we have the identification

3.4.31 $\qquad \phi_1(z, T) = P(0)/P(z - 1).$

If we set $1 - z = \zeta$, then we obtain

3.4.32 $\qquad \phi(1 - \zeta, T) = e^{\Gamma T} \Big/ \left\{ \cosh \varepsilon T + \frac{1}{2}\left(\frac{\Gamma}{\varepsilon} + \frac{\varepsilon}{\Gamma}\right) \sinh \varepsilon T \right\},$

where

3.4.33 $\qquad \varepsilon = (\Gamma^2 + 2\Gamma \bar{I}\zeta)^{1/2}.$

In principle the multi-time generating functions Φ_n can be obtained in a similar way; the computations become excessively complex even for $n = 3$.

Example 3.4.2 Let $\{\Lambda(t)\}$ be a stationary Markov process on a two-point state space $\{\lambda - a, \lambda\}$ where $0 \leqslant a \leqslant \lambda$. The two points can be relabelled as 0 and 1 for convenience. The transition probabilities are specified by

3.4.34 $\qquad Pr\{\Lambda(t + \Delta) = 0 | \Lambda(t) = 1\} = \lambda\Delta + o(\Delta)$

3.4.35 $\qquad Pr\{\Lambda(t + \Delta) = 1 | \Lambda(t) = 0\} = \gamma\Delta + o(\Delta) \quad (\gamma > 0).$

In Example 2.2.2 we have discussed such a Markov chain; we take the results *in toto*:

3.4.36 $\qquad \rho_{ij}(t) = Pr\{\Lambda(t) = i | \Lambda(0) = j\}$

3.4.37 $\qquad \rho_{00}(t) = [\lambda + \gamma\, e^{-(\lambda + \gamma)t}]/(\gamma + \lambda)$

3.4.38 $\qquad \rho_{11}(t) = [\gamma + \lambda\, e^{-(\lambda + \gamma)t}]/(\gamma + \lambda).$

The Cox process generated by $\{\Lambda(t)\}$ can be interpreted as follows. Consider a Poisson process with parameter λ and modify it in the following way. Each event or incidence inhibits the occurrence of further events by reducing the parameter by a positive quantity a. The state corresponding to the value $\lambda - a$ is deemed as ground state below which the intensity parameter cannot drop. In other words, if just prior to the occurrence of an incidence, the parameter had a value $\lambda - a$, it continues to remain so at the same value after occurrence of the incidence. However, 3.4.35 implies that there is a propensity for the parameter to move up to the value at a constant rate γ. In other words, the sojourn time in state corresponding to $\lambda - a$ is exponentially distributed with parameter γ. Viewed merely as the process $\{\Lambda(t)\}$, it alternates between the two states. In particular if we take $a = \lambda$, then the Poisson process is *frozen* for a while; this is a familiar case of a counter with random dead time. The dead and free periods are as indicated in Fig. 3.4.1. The time-interval blocks are independent and exponentially

Fig. 3.4.1 Alternation of $\{\Lambda(t)\}$ process

distributed with parameters γ and λ corresponding to dead and free periods. In the general case, the striped blocks correspond to the Poisson parameter being in the ground state; each striped block is triggered by a Poisson event. The free block is really free from events.

We can obtain the product densities interpreting 3.4.6 and 3.4.7 as the conditioned product densities. The stationary value of the first-order product density is given by

$$3.4.39 \qquad h_1(\,.\,) = E[\Lambda(t)] = \frac{\lambda}{\lambda + \gamma}(\lambda - a) + \frac{\gamma}{\lambda + \gamma} = \left(1 - \frac{a}{\lambda + \gamma}\right)\lambda.$$

In other words, the rate λ is renormalized by the factor $1 - a/(\lambda + \gamma)$. The second-order product density is obtained by using combinatorial arguments for $t_2 > t_1$:

$$h_2(t_1, t_2) = E[\Lambda(t_1)]E[\Lambda(t_2)\,|\,\Lambda(t_1) = \lambda_0]$$

$$3.4.40 \qquad = \sum_{i,j=0}^{1} \pi_i \rho_{j0}(t_2 - t_1)\lambda_i \lambda_j,$$

where π_i are the stationary probabilities of the process $\Lambda(.)$ given by

3.4.41 $\qquad \pi_0 = \lambda/(\lambda + \gamma), \quad \pi_1 = \gamma/(\lambda + \gamma)$

and λ_i's are defined by

$$\lambda_0 = \lambda_{-a}, \quad \lambda_1 = \lambda.$$

Using 3.4.37 and 3.4.38, we obtain finally

$$h_2(t_1, t_2) = \left(1 - \frac{a}{\lambda + \gamma}\right)\lambda\left[\lambda\left(1 - \frac{a}{\lambda + \gamma}\right) - \frac{\gamma a}{\lambda + \gamma}\, e^{-(\lambda + \gamma)|t_2 - t_1|}\right]$$

3.4.42 $\qquad = h_{sty}(|t_1 - t_2|).$

It is to be noted that $h_{sty}(|t|)$ has the maximum value at $t = \infty$ while

3.4.43 $\qquad h_{sty}(0) < (h_1(.))^2,$

exhibiting the inhibitory nature of the process. We note that by construction the point process is a stationary renewal process. The process is easily generalized so that the state space of $\{\Lambda(t)\}$ consists of finitely many points.

Example 3.4.3   Space-charge limited electron stream   This is one of the earliest known examples of a self-inhibited process; for a detailed account of the physics of the problem and full justification the reader may consult Moullins (1938). In view of its importance, we formulate the problem; for a complete solution the reader is referred to Srinivasan (1965, 1986a). In a space-charge limited diode, electron arrivals which normally form a Poisson process with a constant parameter tend to inhibit further arrivals by an intricate dependence on the patterns of arrivals. The process of electron arrivals at the anode can be modelled as a conditioned Poisson process with the parameter $\Lambda(t)$ which is itself a stochastic process subject to the following constraints. Each electron arrival at the anode in the time-interval $(t_i, t_i + dt_i)$ diminishes the rate of arrival $\Lambda(t)$ by $b\,e^{-a(t - t_i)}$ $(t > t_i)$ where a and b are positive constants depending on the physical characteristics of the anode system. To model the process, we assume that $\Lambda(t)$ satisfies the differential relation

3.4.44 $\qquad d\Lambda(t) = -a(\Lambda(t) - \lambda_0) - b\,dN(t),$

where λ_0 is the initial value of $\Lambda(t)$ at $t = 0$ and $N(t)$ is the counting process of the stream of electron arrivals. If $\{t_i\}$ are the epochs of arrivals, then

3.4.45 $\qquad \Lambda(t) = \lambda_0 - b\sum_i \exp -a(t - t_i).$

There is an interesting possibility of the process $\Lambda(t)$ making excursions in the range $(-b, 0]$, during which time no arrivals are possible. In view of this, it is convenient to introduce the process $\Lambda'(t)$, where

3.4.46
$$\Lambda'(t) = \Lambda(t) \quad \Lambda(t) > 0$$
$$= 0 \quad \text{otherwise,}$$

where $\Lambda(t)$ is always taken to be the left-hand limit $\Lambda(t - 0)$. Thus it is the process $\Lambda'(t)$ that is physically significant, and we can identify the space-charge-limited electron stream to be the conditional Poisson process driven by $\Lambda'(t)$. We can analyze the process by observing that $\Lambda(t)$ is Markov for a given initial set-up value λ_0 and then construct a stationary process with stationary equilibrium distribution by receding the time origin to $-\infty$. This will pave the way for arriving at explicit expressions for the product densities of the resulting stationary point process.

Additional notes

The theory of point processes has been studied from many angles in the past. Perhaps the earliest result is due to Campbell (1909) and relates to the response process generated by a series of independent events; this was generalized and an inhibited Poisson process was introduced by Rowland (1936, 1937, 1938) who attempted to obtain an explicit formula for the spectrum of the resulting response process. Around the same time, Von Mises (1936) introduced point events in phase-space and studied the stability properties of certain functionals associated with the point events. Later Rice (1945) studied the properties of stationary point events generated by the zero crossings of the noise processes. In a slightly different context Palm (1943) had attempted to describe the process of arrivals in a telephone network; this was pursued by Khintchine (1955) who laid strong foundations for a mathematical structure of the theory of stationary point processes.

There has been a second line of approach arising from studies in population growth and cosmic ray showers. David Kendall (1949) studied the age-dependent population growth through the study of linear and quadratic cumulant functions. This is an explicit demonstration of handling of evolutionary (non-stationary) population process through the study of point distribution over the continuum, the continuum representing the age parameter of the members of the population. Bhabha (1950) and Ramakrishnan (1950) introduced appropriate measures of the counting process of cosmic ray particles distributed over energy, and developed the theory of moment measures. They explicitly demonstrated how the basic Markov process characterizing the cosmic-ray shower models can be used to obtain Chapman–Kolmogorov type equations for the moment measures. Janossy (1950a) had also attacked the problem indirectly by invoking the Chapman–Kolmogorov equation satisfied by the finite dimensional distributions defined over the energy parameter. The generality of the method was established by Ramakrishnan (1950, 1953, 1958) who showed how the point distribution in different situations can be analysed in a systematic manner. Later Moyal (1962) combined many of these findings and developed a formal, well-knit theory of population point processes defined on an arbitrary set.

A third line of development was motivated primarily by the results of random walk and renewal theory. Renewal theory had its origin in the study of self-renewing aggregates and population growth and was brought to the fore by Feller (1941, 1949) and later by Smith (1954, 1958) and Cox (1962). Cox identified the process as a replacement process in which the points are on the time-axis corresponding to the failure and/or replacement of machines. This led to a generalization of the renewal process to Markov renewal theory and semi-Markov processes. The renewal process provides a convenient starting-point for further generalizations, and the contributions by Kuznetsov and Stratonovic (1956) and McFadden (1962) are in this direction.

There is a different type of generalization of a Poisson process that is very fruitful; this arises when the parameter characterizing the process is itself a stationary positive-valued stochastic process. A simple example of such a process was noted by Cox (see Bartlett 1963) in connection with the modelling of the process of stoppages in a textile loom. Such processes are known as *doubly stochastic Poisson processes* or *Cox processes*. A generalization of such processes leads to self-excitation/inhibition in a natural way and was studied by Srinivasan (1965, 1986a) in the context of models of space-charge limited shot noise. The examples and the method of analysis presented in section 3.4 have been discussed in a more general setting by Srinivasan and Sukavanam (1971, 1972) and Srinivasan (1974b, 1986b).

Multidimensional point processes have also been studied in the literature; in the cascade theory of cosmic-ray showers bivariate point processes in two dimensions were investigated by Ramakrishnan and Srinivasan (1956); Vere-Jones (1970) studied a special process on the surface of the sphere in connection with a model of earthquake.

A further generalization consists in visualizing different types of events;

in fact, the point distribution conceived by Bhabha and Ramakrishnan pertains to the population point process of electrons and photons, and this describes an evolutionary bivariate point process. A formal account of the theory of multivariate point processes is given in a survey by Cox and Lewis (1972). Point processes have also been formulated in non-Euclidean spaces; Dyson and McVoy (1958) have investigated polarization cascades; a detailed account can be found in Srinivasan (1969).

The literature on the subject is rather heavy and lies scattered in different journals, proceedings of conferences and symposia. There is a valuable survey by Daley and Vere-Jones (1972); the volume edited by Lewis (1972) gives an idea of the trend and potential for further growth. The Davidson memorial volume edited by Kendall and Harding (1973, 1974) contains articles dealing with abstract formulation leading to theory of random measures. The following monographs more or less deal with point processes and applications of the theory: Harris (1963), Cox and Lewis (1966), Murthy (1974), Srinivasan (1969, 1974a), Snyder (1975), Kallenberg (1975), Mathes (1978), Cox and Isham (1980), Srinivasan and Subramanian (1980), Sampath and Srinivasan (1977).

4. Population growth process

In this chapter we present some of the techniques necessary to obtain the characteristics of population processes in the context of cavity population photons. We first present a fairly detailed analysis of a birth-and-death process supported by an immigration process. Since the cavity population is subject to detection, as briefly mentioned in Chapter 1, we have brought to focus the emigration process that arises when each individual has a constant probability per unit time of emigrating out of the system. Since explicit solution by direct analytical methods is possible, we have presented the analysis in all its details. In section 4.2 we present the age-dependent population process of the Bellman–Harris type. We then discuss the Kendall process and in particular a special type of age dependency which admits an interesting interpretation in terms of phases. Explicit formulae for the moments and age-dependent product densities are also provided.

4.1 Population model with constant rates: (a) Formulation

We deal with a simple model of population growth of the type discussed in Examples 2.2.5 and 3.2.1. Each individual, independent of other members of the population, has a constant probability λ per unit time of giving birth to another, μ, of death and η of emigration out of the system. In addition there is a constant probability v per unit time of immigration of an individual into the system. Let $X(t)$ be the size of the population at time t and $Y(t)$ the number of individuals who have emigrated over the interval $(0, t)$. At the outset we note that *in the absence of immigration* the population process $\{X(t)\}$ is of the branching type in the sense that the evolution of the population process $X(t)$ generated by co-existing individuals is statistically independent. Accordingly we introduce the following convenient notation:

$$p_1(n, t) = Pr\{X(t) = n \,|\, X(0) = 1, v = 0\}$$

4.1.1
$$p(n, t) = Pr\{X(t) = n \,|\, X(0) = 0, v \neq 0\}$$

$$P_1(n, t) = Pr\{Y(t) = n \,|\, X(0) = 1, v = 0\}$$

$$P(n, t) = Pr\{Y(t) = n \,|\, X(0) = 0, v \neq 0\}$$

We next introduce the generating functions of the above probabilities:

$$g_1(z, t) = E\{z^{X(t)} \mid X(0) = 1, v = 0]$$

4.1.2
$$g(z, t) = E[z^{X(t)} \mid X(0) = 0, v \neq 0]$$
$$G_1(z, t) = E[z^{Y(t)} \mid X(0) = 1, v = 0]$$
$$G(z, t) = E[z^{Y(t)} \mid X(0) = 0, v \neq 0].$$

We now deal with connecting relations between the various probabilities 4.1.1. We first deal with $p(n, t)$ and obtain the consequences of the Chapman–Kolomogorov relation by concentrating our attention on the interval $(0, \Delta)$ for arbitrarily small Δ. There are two mutually exclusive possibilities according as whether an immigration takes place or not. If an immigration takes place, then the process can be conceived as the sum of two independent population processes, one generated by the immigrant and the other being the residue. Thus we have, using the homogeneous nature of the process,

$$p(n, t) = (1 - v\Delta)p(n, t - \Delta)$$

4.1.3
$$+ \Delta \sum_{n=0}^{n} vp_1(n_1, t - \Delta)p(n - n_1, t - \Delta) + o(\Delta).$$

Taking the limit as $\Delta \to 0$, we obtain

4.1.4
$$\frac{\mathrm{d}}{\mathrm{d}t} p(n, t) = -vp(n, t) + \sum_{n_1=0}^{n} p_1(n, t)p(n - n_1, t);$$

ir equivalently in terms of the corresponding generating functions, we have

4.1.5
$$\frac{\partial}{\partial t} g(z, t) = -vg(z, t) + vg(z, t)g_1(z, t)$$

whose solution can be written down using the initial condition $g(z, 0) = 1$. Thus we have

4.1.6
$$g(z, t) = \exp -v \int_0^t [1 - g_1(z, \tau)]\,\mathrm{d}\tau.$$

There is an identical relationship between the functions $P_1(n, t)$ and $P(n, t)$. The equations are obtained by replacing the functional symbols p_1, p, g_1 and g by the corresponding capital letters P_1, P, G_1 and G. Thus it is sufficient if we obtain explicit solutions for $p_1(n, t)$ and $P_1(n, t)$.

4.1 (b) Differential equations governing the generating functions and their solutions

Next we write down the backward equation satisfied by $g_1(z, t)$ directly. As before, we deal with the possibilities over the sub-interval $(0, \Delta)$ of the interval $(0, t)$. The single individual may die or emigrate or produce another individual over the interval $(0, \Delta)$ with respective probabilities $\mu\Delta$, $\eta\Delta$, and $\lambda\Delta$, or it may survive these possibilities with residual probability $1 - (\lambda + \mu + \eta)\Delta + o(\Delta)$. Thus we have

$$g_1(z, t) = [1 - (\lambda + \mu + \eta)\Delta]g_1(z, t - \Delta) + (\mu + \eta)\Delta$$

4.1.7
$$+ \lambda\Delta[g_1(z, t - \Delta)]^2 + o(\Delta)$$

which in turn leads to

4.1.8
$$\frac{\partial g_1(z, t)}{\partial t} = -(\lambda + \mu + \eta)g_1(z, t) + \mu + \eta + \lambda[g_1(z, t)]^2$$

with the initial condition

4.1.9
$$g_1(z, 0) = z.$$

In an exactly similar way, we find that the function $G_1(z, t)$ satisfies the equation

4.1.10
$$\frac{\partial G_1(z, t)}{\partial t} = -(\lambda + \mu + \eta)G_1(z, t) + \mu + \eta z + \lambda[G_1(z, t)]^2$$

with the initial condition

4.1.11
$$G_1(z, 0) = 1.$$

Equations 4.1.10 and 4.1.8 are different and hence have to be solved separately. To save computational labour we can deal with the joint distribution of $X(t)$ and $Y(t)$. Accordingly we define $H_i(z_1, z_2, t)$ $(i = 1, 2, \ldots)$ and $H(z_1, z_2, t)$ by

4.1.12
$$H_i(z_1, z_2, t) = E[z_1^{X(t)} z_2^{Y(t)} \,|\, X(0) = i, v = 0]$$

4.1.13
$$H(z_1, z_2, t) = E[z_1^{X(t)} z_2^{Y(t)} \,|\, X(0) = 0, v \neq 0].$$

As before, we can show that H is related to H_1 by

4.1.14
$$H(z_1, z_2, t) = \exp -v \int_0^t [1 - H_1(z_1, z_2, \tau)]\,d\tau.$$

Next we note that since

4.1.15 $\qquad H_i(z_1, z_2, t) = [H_1(z_1, z_2, t)]^i \quad i = 1, 2, \ldots$

it is sufficient to deal with $H_1(z_1, z_2, t)$ which, in turn, satisfies the equation

$$\frac{\partial H_1(z_1, z_2, t)}{\partial t} = -(\lambda + \mu + \eta)H_1(z_1, z_2, t)$$

4.1.16 $\qquad\qquad\qquad + \mu + \eta z_2 + \lambda[H_1(z_1, z_2, t)]^2$

with the initial condition

4.1.17 $\qquad H_1(z_1, z_2, 0) = z_1.$

Equation 4.1.16 can be solved directly; we obtain after some routine calculations,

4.1.18 $\qquad H_1(z_1, z_2, t) = \xi_- + \dfrac{\Delta(z_2)}{1 - \left(1 - \dfrac{\Delta(z_2)}{z_1 - \xi_-}\right)\exp\{\lambda t \Delta(z_2)\}}$

where

4.1.19 $\qquad 2\xi_{\pm} = \dfrac{(\lambda + \mu + \eta)}{\lambda} \pm \Delta(z_2)$

4.1.20 $\qquad \lambda\Delta(z_2) = [(\lambda + \mu + \eta)^2 - 4\lambda(\mu + \eta z_2)]^{1/2}.$

From 4.1.18 we obtain $G_1(z, t)$ and $g_1(z, t)$:

4.1.21 $\qquad G_1(z, t) = H_1(1, z, t)$

$\qquad\qquad g_1(z, t) = H_1(z, 1, t)$

4.1.22 $\qquad\qquad = 1 + \dfrac{(\mu + \eta - \lambda)\,e^{-(\mu + \eta - \lambda)t}(z - 1)}{\mu + \eta - \lambda z + \lambda(z - 1)\,e^{-(\mu + \eta - \lambda)t}}.$

Using 4.1.6 we finally obtain

4.1.23 $\qquad g(z, t) = [(\mu + \eta - \lambda)/\{\mu + \eta - \lambda z + \lambda(z - 1)\,e^{-(\mu + \eta - \lambda)t}\}]^{\nu/\lambda}$

4.1.24 $\qquad G(z, t) = e^{-\nu t(1 - \xi_-)}\left[\dfrac{\Delta(z)}{(1 - \xi_-)\,e^{-\lambda t \Delta(z)} - (1 - \xi_+)}\right]^{\nu/\lambda}.$

4.1 (c) Steady-state characteristics

If we assume $\mu + \eta > \lambda$, then the population process $\{X(t)\}$ has a non-trivial limiting (equilibrium) distribution. Denoting the corresponding generating function by a subscript E, we obtain

4.1.25 $$g_E(x) = \lim_{t \to \infty} g(z, t) = \left[\frac{\mu + \eta - \lambda}{\mu + \eta - \lambda z} \right]^{v/\lambda}.$$

Thus the *equilibrium population has a negative binomial distribution*.

There is another process of interest; we define the process $W(t, T)$ by

$$W(t, T) = Y(t + T) - Y(t),$$

corresponding to the size of the emigrants over the time-interval $(t, t + T)$, particularly when t is very large. We show first that the generating function of $W(t, T)$ has a limit as $t \to \infty$. First we notice

$$E[z^{W(t,T)} \mid X(0) = m]$$

$$= \sum Pr\{X(t) = n \mid X(0) = m\} E[z^{W(t,T)} \mid X(t) = n]$$

$$= \sum Pr\{X(t) = n \mid X(0) = m\} [G_1(z, T)]^n G(z, T)$$

4.1.26 $$= G(z, T)g(G_1(z, T), t)[g_1(G_1(z, T), t)]^m.$$

The r.h.s. of 4.1.26 has a limit as $t \to \infty$ and hence we have

4.1.27 $$\lim_{t \to \infty} E[z^{W(t,T)} \mid X(0) = m] = G(z, T)g_E(G_1(z, T)).$$

From the above result we conclude that we can generate a stationary point process of emigration if we recede the time-origin to $-\infty$. Of course 4.1.27 relates only to the behaviour of the first-order distribution; however, the result can be extended to the simultaneous distribution of the number of emigrants over many time-intervals.

4.1 (d) Point process of emigrations: product densities

Next we study the point process of emigrations. We have seen in Example 3.3.3 that the births and deaths form a point process and that the process is stationary provided we recede the time-origin to minus infinity. If we recede the time-origin to minus infinity, we note that we obtain an equilibrium distribution of population size with equilibrium probability of the size specified by 4.1.25. It may be interesting to obtain the product densities of the stationary point process. We first introduce the conditional product-density of emigrations by

4.1.28 $$_1h_1(t) = \lim_{\Delta \to 0} Pr\{Y(t_1 + \Delta) - Y(t_1) = 1 \mid X(0) = 1, v = 0\}/\Delta.$$

It is possible to set up a differential equation for $_I h_1^1(t)$ as in Example 3.3.3. However, we can obtain $_I h_1^1(t)$ directly by appealing to the definition 4.1.28. We have, in fact,

$$Pr\{Y(t+\Delta) - Y(t) = 1 \,|\, X(0) = 1, v = 0\}$$

$$= \sum_{n=0}^{\infty} \eta n \Delta Pr\{X(t) = n \,|\, X(0) = 1, v = 0\} + o(\Delta)$$

$$4.1.29 \qquad = \eta \Delta \left. \frac{\partial g_1(z, t)}{\partial z} \right|_{z=1} + o(\Delta).$$

$$4.1.30 \qquad _I h_1(t) = \eta \, e^{-(\mu + \eta - \lambda)t}.$$

Next we define the unconditional product density of degree 1 of emigrations:

$$4.1.31 \qquad h_1(t) = \lim_{\Delta \to 0} Pr\{Y(t+\Delta) - Y(t) = 1\}/\Delta.$$

Again using the definition on the r.h.s. of the above relation, we find

$$4.1.32 \qquad h_1(t) = \eta \left. \frac{dg_E(z)}{dz} \right|_{z=1}$$

since we have an equilibrium distribution of population at any finite time-point. Using the solution 4.1.25 we finally have

$$4.1.33 \qquad h_1(t) = \eta v/(\mu + \eta - \lambda).$$

There is also another conditional product density, $_0 h_1(t)$ (similar to the one introduced in Example 3.3.3) which will be useful in the determination of the second-order product density and can be defined by

$$4.1.34 \qquad _0 h_1(t) = \lim_{\Delta \to 0} Pr\{Y(t+\Delta) - Y(t) = 1 \,|\, X(0) = 0, v \neq 0\}/\Delta.$$

The function $_0 h_1(t)$ can be obtained by writing the backward differential relation

$$_0 h_1(t) = (1 - v\Delta)_0 h_1(t - \Delta)$$

$$4.1.35 \qquad\qquad + v\Delta[_0 h_1(t - \Delta) + _I h_1(t - \Delta)] + o(\Delta)$$

which leads to

$$4.1.36 \qquad \frac{d}{dt} \, _0 h_1(t) = v \,_I h_1(t).$$

Using 4.1.30, we obtain

$$4.1.37 \qquad {}_0h_1(t) = \frac{v\eta}{\mu + \eta - \lambda}(1 - e^{-(\mu + \eta - \lambda)t}).$$

If we have stationary equilibrium condition at the origin, then it is easy to see that the following relation

$$4.1.38 \qquad h_1(t) = \lim_{\tau \to \infty} {}_0h_1(\tau) = v\eta/(\mu + \eta - \lambda)$$

holds good, consistent with 4.1.33.

Next we deal with the second-order product density. We will assume stationary equilibrium condition at the origin. Then it is easy to see that the product density of degree two defined by

$$h_2(t_1, t_2) = \lim_{\Delta, \Delta' \to 0} Pr\{Y(t_1 + \Delta) - Y(t_1) = 1,$$

$$4.1.39 \qquad\qquad Y(t_2 + \Delta') - Y(t_2) = 1\}/\Delta\Delta'$$

is a function of $|t_2 - t_1|$ only. If we denote the resulting function by $h_{sty}(t)$ $(t > 0)$, then it can be interpreted as the product density corresponding to two emigrations separated by an interval of length t. We note then that the emigration that takes place at "epoch" t may be either

   (i) from the population generated by an individual present at the new time-origin, or

   (ii) from the population generated by an immigrant entering the system between 0 and t.

Thus we have

$$4.1.40 \qquad h_{sty}(t) = \sum_{n=0}^{\infty} \eta n Pr\{X(0-) = n\}\left[(n-1)_I h_1(t) + {}_0h_1(t)\right].$$

Observing that the probability distribution at $0-$ corresponds to the equilibrium distribution, we have

$$4.1.41 \qquad h_{sty}(t) = \left[\eta \frac{d^2 g_E}{dz^2}\bigg|_{z=1}\right]_I h_1(t) + \left[\eta \frac{d g_E}{dz}\bigg|_{z=1}\right]_0 h_1(t).$$

Using 4.1.37 and 4.1.25, we have

$$4.1.42 \qquad h_{sty}(t) = \eta^2 v[v + \lambda e^{-(\mu + \eta - \lambda)t}]/(\mu + \eta - \lambda)^2.$$

If we define a measure of bunching $\mathscr{B}$ of emigrations defined by

$$4.1.43 \qquad \mathscr{B} = h_{sty}(0)/h_{sty}(\infty),$$

we have in this model of population

4.1.44 $\qquad \mathscr{B} = 1 + \lambda/\nu,$

showing that the epochs of emigration have a tendency to cluster.

Higher-order product densities can be obtained in a similar fashion. We finally observe that although the process $\{Y(t)\}$ is non-Markov, the two-dimensional process $\{X(t), Y(t)\}$ is indeed Markov. In the analysis presented so far we have made use of this property by using appropriate conditioned probabilities. For instance, in equation 4.1.40, the dependency on the initial conditioning is apparent.

4.2 Age-dependent population growth: Bellman–Harris process

So far we have discussed the characteristics of population growth when each individual has a constant rate (probability per unit time) of giving birth to another; besides, each individual is also assumed to have a constant rate of death/emigration. If the assumption of constancy of rates is relaxed, most of the analysis presented earlier becomes inapplicable. This is due to the process turning non-Markov. At any rate the constancy of the birth rate per individual is an over-simplification in many models, and hence some kind of an age-dependence on the birth rate is generally postulated so that the model can be used in a wide variety of physical/biological situations. We first consider a model in which an individual born at $t = 0$ has a random life-span (sometimes called *generation time*) L with the probability density function $f(.)$. At the end of the life span, it is replaced by a random number i of similar individuals with age 0 with probability p_i, the probability p_i being independent of the age of the individual as well as the time-point. Death is included by permitting the value $i = 0$ with $p_0 > 0$. Such age-dependent growth processes naturally arise as models for the growth of bacterial colonies in certain phases or species that grow by fission; neutron multiplication under certain conditions can also be modelled by such an age-dependent growth process. The process is named after Bellman and Harris who provided a method of analysis of such problems.

Let $X(t)$ be the size of the population at time t. In general, $\{X(t)\}$ is non-Markov; of course when $f(t)$ corresponds to the exponential distribution, then the process is Markov. At any time t, each individual is characterized by its age and thus we have a point process for each t, the point process describing the distribution of the individuals over the age parameter. Thus we can deal with the product densities of the point process; since for each t

we have a hierarchy of product densities, it may be necessary to study the behaviour of the product densities as t varies. The description of the population in terms of the product densities will bring out the age structure of the members of the population; however, this will be a very difficult task for a general type of age dependency. Nevertheless questions relating to the size of the population without reference to age can be settled, and this we shall proceed to do.

Let $\pi(n, t)$ be the probability mass function of $X(t)$ due to one individual of age 0 at $t = 0$. Then it follows that $\pi(n, t)$ satisfies

$$4.2.1 \qquad \pi(n, t) = \bar{F}(t)\delta_{n1} + \sum_i p_i \int_0^t f(\tau)\pi^{(i)}(n, t - \tau)d\tau,$$

where $\bar{F}(t)$ is the surviver function corresponding to $f(t)$ and $\pi^{(i)}(n, t)$ is the i-fold convolute of the probabilities $\pi(n, t)$. The above equation is obtained by arguing that the individual's life span can either terminate or not, over the interval $(0, t)$, the probability for the latter event being $\bar{F}(t)$. If the life span terminates, then with probability p_i there are i individuals at the termination epoch each of which generates a population over the residual part of the interval. If $g(z, t)$ is the probability-generating function of $\pi(n, t)$, then the above equation reduces to

$$4.2.2 \qquad g(z, t) = z\bar{F}(t) + \int_0^t f(\tau)q(g(z, t - \tau))d\tau,$$

where $q(.)$ is the generating function of the probabilities $\{p_i\}$. Equation 4.2.2 is not capable of explicit solution in general. However, the moments of the size of the population can be obtained by differentiating both sides of g w.r.t. z.

There is a special case of considerable biological interest when $f(t)$ is given by

$$4.2.3 \qquad f(t) = e^{-\alpha t}\frac{(\alpha t)^n}{n!}\alpha,$$

where α is any positive parameter. When $n = 0$, the case reduces to the exponentially distributed life span; in such a case, if we choose

$$4.2.4 \qquad \alpha = \lambda + \mu, \quad p_0 = \mu/(\lambda + \mu), \quad p_2 = \lambda/(\lambda + \mu), \qquad p_i = 0, \quad i \neq 0, 2$$

then the model reduces to the birth-and-death process discussed in section 4.1. When $n > 0$, each individual can be assumed to go through $(n + 1)$ phases whose spans are independent and identically distributed with the common exponential distribution with parameter α. If we denote the size of the population of individuals in the ith phase by $X_i(t)$ $(i = 1, 2, \ldots, n+1)$ and retain $X(t)$ to denote the total population size, we can introduce the generating function $g_i(z, t)$, where

4.2.5 $\qquad g_i(z, t) = E[z^{X_i(t)} \,|\, X_i(0) = 1 = X(0)],$

then we obtain by the backward differential relation,

4.2.6 $\qquad \dfrac{\partial g_i(z, t)}{\partial t} = -\alpha g_i(z, t) + \alpha g_{i+1}(z, t) \quad 1 \leqslant i \leqslant n$

4.2.7 $\qquad \dfrac{\partial g_{n+1}(z, t)}{\partial t} = -\alpha g_{n+1}(z, t) + \alpha q(g_1(z, t)),$

with the initial condition $g_i(z, 0) = z$. If on the other hand we want a more detailed description of the population, then we can introduce the function $H_i(z_1, z_2 \cdots z_n, z_{n+1}, t)$ where

$$H_i(z_1, z_2, \ldots, z_n, z_{n+1}, t)$$

4.2.8 $$= E\left[\prod_{j=1}^{n+1} z_j^{X_j(t)} \,\Big|\, X_i(0) = X(0) = 1\right]$$

and again deal with the backward differential relation, which leads to equations 4.2.6 and 4.2.7 (if we replace g_i by H_i) with the initial condition

4.2.9 $\qquad H_i(z_1, z_2, \ldots, z_{n+1}, 0) = z_i.$

Returning to the general model, we note that in the model of growth described above, the individuals are assumed to be of the same type. It is possible to modify equation 4.2.2 to include cases in which different types of individuals are produced at the time of fission.

4.3 Age-dependent Kendall process: method of phases

In the Bellman–Harris process discussed, the age-dependence of the growth process has been incorporated on a renewal basis; this may not be applicable to situations where the parent has a fertile period during which it gives rise to offsprings. Let us consider the population process introduced in section 4.1 and remove the constancy of the rate of birth and replace it by an age-dependent function $\lambda(x)$, where x is the age of the individual in question. Let us take out of our preview the immigration and emigration process, so that we have a pure birth- and-death process. David Kendall studies such processes extensively with special reference to questions of age-structure of the population; the process is known as the Kendall process. For clarity we restate the assumptions of the model:

  (i) the sub-populations generated by co-existing cells* develop in complete independence of each other;

  (ii) given a cell of age x at time t, the probability that it produces another cell of age zero in the time-interval $(t, t + \Delta)$ is given by $\lambda(x)\Delta + o(\Delta)$.

  (iii) given a cell of age x at time t, the probability that it dies in the interval $(t, t + \Delta)$ is given by $\mu(x)\Delta + o(\Delta)$.

It is to be specially noted that the birth-and-death rates do not depend on t, the epoch of its existence. In view of the independence of sub-populations, it is sufficient if we study the population generated by a single individual of given age x_0 initially at $t = 0$. At any arbitrary time $t > 0$, we have a point process corresponding to the distribution of cells in age. David Kendall (1949) developed a theory of cumulant functions and cumulant-generating functionals in order to characterize the age-structure of the population. The problem becomes pretty complex and depends on the solution of Fredholm type integral equations. To illustrate the point, we analyze the problem by dealing with the limited question of distribution of the total size of the population irrespective of the age. If $X(t)$ is the size at time t of the population generated by a single individual of age x_0, we note that the generating function $g(z, t, x_0)$ defined by

$$4.3.1 \qquad g(z, t, x_0) = E[z^{X(t)} \mid X(0) = 1 = N(dx_0)],$$

where $N(.)$ is the initial point distribution of cells over age, satisfies the backward differential relation

$$g(z, t, x_0) = [1 - \lambda(x_0)\, dt - \mu(x_0)\, dt]g(z, t - dt, x_0 + dt)$$

$$+ \lambda(x_0)\, dt g(z, t - dt, x_0 + dt)g(z, t - dt, 0)$$

$$4.3.2 \qquad\qquad + \mu(x_0)\, dt + o(dt)$$

which leads to

$$\left(\frac{\partial}{\partial t} - \frac{\partial}{\partial x_0}\right)g(z, t, x_0)$$

$$= -[\lambda(x_0) + \mu(x_0)]g(z, t, x_0)$$

$$4.3.3 \qquad\qquad + \lambda(x_0)g(z, t, x_0)g(z, t, 0) + \mu(x_0)$$

---

* For brevity we shall use the word cell instead of individual.

with the initial condition

4.3.4 $\qquad g(z, 0, x_0) = z.$

The above equation is indeed difficult to deal with, and only in some special cases it is possible to obtain the first few moments of $X(t)$.

It is possible to deal with the process for some special choice of $\lambda(x_0)$. The constancy of the rate of birth is rather unphysical in some situations; it may be desirable to have zero to start with, the rate increasing with age up to a certain stage and falling off to zero ultimately. An appropriate choice of $\lambda(x)$ meeting this requirement is given by

4.3.5 $\qquad \lambda(x) = e^{-\lambda x} \dfrac{(\lambda x)^{n-1}}{(n-1)} \alpha,$

where α has the dimension of reciprocal of time to make $\lambda(x)$ dimensionally correct. For simplicity we take $\mu(x)$ to be a constant equal to μ, although the arguments that follow are equally valid for any arbitrary $\mu(x)$.

The rate $\lambda(x)$ as defined by 4.3.5 has the following interpretation. We note that the rate itself is conditional upon its survival up to the epoch under consideration. The conditional life span x can be interpreted to be the sum of $(n + 1)$ random variables, the random variables corresponding to the life spans of phases. The spans corresponding to the first n phases are independent and exponentially distributed with the same parameter λ; the span of the last phase is defined as the residue. The birth rate is a constant equal to α if the cell is in phase n, the birth rate being zero if it is in any of the other phases. In other words, the cell has to attain a maturity to produce another cell and this is attained by the cell moving forward step by step to the nth phase. The passage to the $(n + 1)$th phase renders the cell sterile. We now establish that *this interpretation is consistent with the choice $\lambda(x)$* implied by 4.3.5.

We note that the conditional probability $\beta_k(x)$, given the age x, of a cell being found in phase k is given by

4.3.6 $\qquad \beta_k(x) = e^{-\lambda x} \dfrac{(\lambda x)^{k-1}}{(k-1)!} \qquad (k = 1, 2, \ldots, n)$

4.3.7 $\qquad \beta_{n+1}(x) = 1 - \sum_{k=1}^{n} \beta_k(x).$

From the above probabilities we can compute the probability, given the age x, of a cell giving birth to another. Thus we have

4.3.8 $\qquad \lambda(x) = \beta_n(x)\alpha,$

which is consistent with 4.3.5.

It is to be noted that the conditional life span of the $(n+1)$th phase may be infinite; however, the cell will ultimately die with probability one. It should also be noted that the modified process is different from the Kendall process inasmuch as the births in two disjoint intervals due to the same individual are longer independent. We call the process the *modified Kendall process*. The marginal birth probabilities in the modified process are the same as in the original Kendall process, and the model specified by 4.3.5 and interpreted as above is fairly versatile, since several ramifications are possible keeping the phases intact. For instance we can assume differential death rates by taking $\mu(x) = \mu_k$ if the cell is in phase k. Such models have been investigated in the past and it may be fruitful to study them. It is also possible to generalize the model so that non-zero birth rates may be assigned to more than one phase. A further generalization is possible in which the movement of the cell with respect to time in phases need not be to the next consecutive phase: for instance, a movement regulated by a Markov chain over the $n+1$ phases (notice that $n+1$ is some kind of an absorbing state) may be an apt description in certain types of bacterial growth. Since the formation of protein molecules is essentially by synthesis, this type of modelling may account for the usual delay that is observed in the movement of enzymes. In any case the interpretation of movement through phases is fairly versatile for modelling diverse phenomena.

We now confine our attention to a model of population growth corresponding to $\lambda(x)$ specified by 4.3.5 when $n = 2$, so that we have a three-phase model and $\lambda(x)$ is now defined by

4.3.9 $\qquad \lambda(x) = \alpha\lambda e^{-\lambda x}x.$

We take $\mu(x) = $ a constant $= \mu$ in all the phases and proceed to characterize the age-structure of the population. Let $X_i(t)$ be the number of cells in phase i at time t and $X(t)$ the total number of cells at time t. We introduce the generating function $g_i(z_1, z_2, t)$ by

4.3.10 $\qquad g_i(z_1, z_2, t) = E\left[z_1^{X_1(t)} z_2^{X_2(t)} \,\middle|\, X_i(0) = 1 = X(0)\right]$

and use the backward differential relation to obtain

4.3.11 $\qquad \dfrac{\partial g_1(z_1, z_2, t)}{\partial t} = -(\lambda + \mu)g_1(z_1, z_2, t) + \lambda g_2(z_1, z_2, t) + \mu$

$$\frac{\partial g_2(z_1, z_2, t)}{\partial t} = -(\lambda + \mu + \alpha)g_2(z_1, z_2, t)$$

$$4.3.12 \qquad\qquad + \alpha g_2(z_1, z_2, t)g_1(z_1, z_2, t) + \mu + \lambda,$$

with the initial conditions

$$4.3.13 \qquad g_i(z_1, z_2, z) = z_i \quad i = 1, 2.$$

Although it may not be possible to solve for g_1 and g_2 explicitly, the first few factorial moments and correlations (between phases) can be obtained.

4.4 Kendall process: moments and age-structure

To obtain the moments, we introduce the functions $a_i^j(t)$, $b_i^{jk}(t)$ $(i, t, k = 1, 2)$ by

$$4.4.1 \qquad a_k^j(t) = \left.\frac{\partial g_i}{\partial z_j}\right|_{z_1 = z_2 = 1}$$

$$4.4.2 \qquad b_i^{jk}(t) = \left.\frac{\partial^2 g_i}{\partial z_j \partial z_k}\right|_{z_1 = z_2 = 1}.$$

We then differentiate both sides of 4.3.11 and 4.3.12 to otbain, for $j = 1, 2$,

$$4.4.3 \qquad \frac{da_1^j(t)}{dt} = -(\lambda + \mu)a_1^j(t) + \lambda a_2^j(t)$$

$$4.4.4 \qquad \frac{da_2^j(t)}{dt} = -(\lambda + \mu)a_2^j(t) + \alpha a_1^j(t)$$

with the initial conditions

$$4.4.5 \qquad a_1^1(0) = a_2^2(0) = 1, \quad a_2^1(0) = a_1^2(0) = 0.$$

The solution of these equations is easily obtained:

$$4.4.6 \qquad a_1^1(t) = a_2^2(t) = e^{-(\lambda + \mu)t} \cosh \sqrt{\lambda \alpha t}$$

$$4.4.7 \qquad \frac{\alpha}{\lambda} a_1^2(t) = a_2^1(t) = \sqrt{\frac{\alpha}{\lambda}} e^{-(\lambda + \mu)t} \sinh \sqrt{\lambda \alpha t}.$$

Next we consider the b-functions which satisfy the following equations.

$$4.4.8 \qquad \frac{d}{dt} b_1^{jk}(t) = -(\lambda + \mu)b_1^{jk}(t) + \lambda b_2^{jk}(t)$$

4.4.9 $\quad \dfrac{\mathrm{d}}{\mathrm{d}t} b_2^{jk}(t) = -(\lambda + \mu)b_2^{jk}(t) + \alpha b_1^{jk}(t) + \alpha[a_1^j(t)a_2^k(t) + a_1^k(t)a_2^j(t)]$

with the initial conditions

4.4.10 $\quad b_i^{jk}(0) = 0 \quad i,j,k = 1,2.$

If we make use of the relations 4.4.6 and 4.4.7, we find

4.4.11 $\quad \lambda b_i^{11}(t) = \alpha b_i^{22}(t) \quad i = 1,2.$

Solving the set of equations 4.4.8 and 4.4.9 for $j = k = 1, i = 1,2$ and $j = 1$, $k = 2, i = 1,2$, we obtain

4.4.12 $\quad b_1^{11}(t) = \alpha \sqrt{\lambda\alpha} \left[\dfrac{p(t)}{D_1} - \dfrac{q(t)}{D_2} + \dfrac{[p(t)]^2}{2D_3} - \dfrac{[q(t)]^2}{2D_4} \right]$

$$b_2^{11}(t) = \alpha^2 \left[\dfrac{p(t)}{D_1} + \dfrac{q(t)}{D_2} \right]$$

$$-\dfrac{\alpha}{2}\sqrt{\dfrac{\alpha}{\lambda}}(\lambda + \mu)\left\{ \dfrac{[p(t)]^2}{D_3} - \dfrac{[q(t)]^2}{D_4} \right\}$$

4.4.13 $\qquad\qquad + \alpha^2 \left\{ \dfrac{[p(t)]^2}{D_3} + \dfrac{[q(t)]^2}{D_4} \right\}$

where

$$p(t) = e^{-(\lambda + \mu - \sqrt{\lambda\alpha})t}, \quad q(t) = e^{-(\lambda + \mu + \sqrt{\lambda\alpha})t}$$

$$D_1 = (\lambda + \mu + \sqrt{\lambda\alpha})^2 - 4\alpha\lambda, \quad D_2 = (\lambda + \mu - \sqrt{\lambda\alpha})^2 - 4\alpha\lambda,$$

4.4.14 $\quad D_3 = (\lambda + \mu - 2\sqrt{\lambda\alpha})^2 - \alpha\lambda, \quad D_4 = (\lambda + \mu + 2\sqrt{\lambda\alpha})^2 - \alpha\lambda.$

The other b-functions are given by

4.4.15 $\quad b_1^{12}(t) = \dfrac{1}{2}b_2^{22}(t) + \dfrac{1}{2}\left(1 + \dfrac{\mu}{\lambda}\right)b_1^{22}(t)$

4.4.16 $\quad b_2^{12}(t) = a_2^1(t) + sb_1^{11}(t) - \left(1 + \dfrac{\mu}{\lambda}\right)b_1^{12}(t).$

We next proceed to discuss the age-structure of the population. Normally we study population generated by an individual of age x_0 at $t = 0$; thus we have a typical population point process at any arbitrary time-instant t with the members of the population distributed over the age parameter which is now confined to the interval $[0, x_0 + t]$. If we denote by

$N(x, t)$ the random variable representing the size, at time t, of the population of individuals with age $\leqslant x$, the typical product densities are conditional ones defined by

$$h_1(x, t \mid x_0)$$

4.4.17
$$= \lim_{\Delta \to 0} Pr\{N(x + \Delta, t) - N(x, t) = 1 \mid N_0(A \mid x) = \delta(A \mid x_0)\}/\Delta,$$

where $N_0(. \mid x)$ is the counting measure (at $t = 0$) introduced in section 3.1. Higher-order product densities are defined in a similar manner. It is much easier to obtain the product densities if we adopt the phase-approach. Accordingly we define product densities conditioned differently. For instance, we can introduce the first-order conditional product densities by

$$h_1^i(x, t) = \lim_{\Delta \to 0} Pr\{N(x + \Delta, t) - N(x, t)$$

4.4.18
$$= 1 \mid X_i(0) = X(0) = 1\}/\Delta \quad (i = 1, 2).$$

If we take $x \leqslant t$, then by elementary combinatorial arguments we have

4.4.19 $$h_1^i(x, t) = a_i^2(t - x)\alpha\, e^{-\mu x} \quad (i = 1, 2)$$

4.4.20 $$h_1^3(x, t) = 0.$$

The product density defined by 4.4.17 is now given by

4.4.21 $$h_1(x, t \mid x_0) = e^{-\lambda x_0} h_1^1(x, t) + \lambda x_0 e^{-\lambda x_0} h_1^2(x, t) \quad x \leqslant t$$

4.4.22 $$h_1(x, t \mid x_0) = e^{-\mu x_0} \delta(x - x_0 - t) \quad x > t,$$

where $\delta(.)$ is the Dirac delta function. Using 4.4.19, 4.4.6 and 4.4.7, we can rewrite 4.4.21 as

4.4.23 $$h_1(x, t \mid x_0) = e^{-\lambda x_0 - \mu x - (\lambda + \mu)(t - x)}$$
$$\times \{\sinh \sqrt{\alpha\lambda}(t - x) + x_0\sqrt{\alpha\lambda} \cosh \sqrt{\alpha\lambda}(t - x)\}\sqrt{\alpha\lambda}.$$

To obtain the second-order product densities, let us, for definiteness, assume $y < x < t$. At the outset we note that if $h_2^i(x, y, t)$ is the product density of degree 2, then the density can be obtained by using as reference point the epoch of birth corresponding to the older of the two cells. Thus the epoch $t - x_0$ just prior to the birth of a cell is the convenient reference time-point. At this epoch the two cells in question may have a common ancestor or two different ancestors. Denoting the corresponding contributions to the product density respectively as P_1 and P_2, we have

4.4.24 $$h_2^i(x, y, t) = P_1 + P_2,$$

where the contribution P_2 can be evaluated directly:

$$P_2 = E[X_2(t-x)X_1(t-x) \mid X_i(0) = X(0) = 1]\alpha\, e^{-\mu x}h_1^1(y,x)$$

$$+ E[X_2(t-x)\{X_2(t-x)-1\} \mid X_i(0) = X(0) = 1]\alpha\, e^{-\mu x}h_1^2(y,x)$$

$$\text{4.4.25} \qquad = \sum_{j=1}^{2} \alpha b_i^{j2}(t-x)\, e^{-\mu x}h_1^j(y,x).$$

To obtain the contribution P_1, we pursue the population tree a little after the epoch $t-x-0$. The cell to be produced later may be produced in the tree generated by the common ancestor present at the epoch $t-x-0$ or in the tree generated by the cell that is produced just after the epoch $t-x-0$. Taking these two mutually exclusive events into account, we have

$$P_1 = E[X_2(t-x) \mid X_i(0) = 1 = x(0)]\alpha$$

$$\times e^{-\mu x}h_1^2(y,x) + K(y,x)$$

$$\text{4.4.26} \qquad = \alpha a_i^2(t-x)\{e^{-\mu x}h_1^2(y,x) + K(y,x)\},$$

where $K(y,x)$ is a mixed product density and $K(y,x)\,dy$ represents the probability that there is a cell of age in $(y, y+dy)$ produced in a population tree generated by the primary conditional upon the primary being born initially. The function $K(y,x)$ can be readily obtained using typical combinatorial arguments:

$$K(y,x) = \alpha\lambda(x-y)\, e^{-\lambda(x-y)-\mu_{(x+y)}}$$

$$\text{4.4.27} \qquad\qquad + \alpha\lambda e^{-\mu x} \int_0^{x-y} e^{-\lambda v}v h_1^1(y, x-v)\,dv.$$

The corresponding product density in the age-dependent version follows by a formula similar to 4.4.21.

It should be noted that the cells in the third phase play practically no role in the evolution process. In the product density description we have taken the possibility of the cell in question being in phase 3 indirectly. This is because we have chosen an asymmetric combination of specifying the age of the cell in question and the phase of the primary cell that generates the population. A more detailed version can be obtained by resorting to a product density description in which the phase is also included. We will illustrate this point in Chapter 8 where we will have an occasion that demands a more detailed description.

Additional notes

Stochastic population growth has a long history, starting from the pioneering contributions by Watson and Galton (1874) dealing with the problem of extinction of surnames. The formulation of population growth problems in terms of probability is perhaps due to Kolmogorov and Dmitriev (1947). The regenerative integral equation in branching processes was first explicitly employed by Bellman and Harris (1948) who showed how age-dependent population evolution can be analyzed. Janossy (1950b) had also introduced, quite independently, in an entirely different context of cosmic ray showers, the method of regeneration by which energy-dependent cascade phenomena can be analyzed. An excellent account of the theory of branching processes is given by Harris (1963) himself. A survey of stochastic population growth was presented by David Kendall (1949) in a symposium highlighting the advances in the theory of stochastic processes. The emigration process and its emergence as a stationary point process when the population is maintained in equilibrium has somehow failed to attract attention; there are only very casual references in the discussion following Kendall's paper. There is also a survey article by Kendall (1966) dealing with subsequent developments and presenting an overview of the subject. The following are some of the monographs that deal with branching processes, particularly with emphasis on limiting behaviour: Athreya and Ney (1972), Mode (1971), Jaggers (1975). In recent times, there are investigations dealing with population growth subject to the Poisson type of disasters; however, the emphasis is on the statistical characteristics of the main population rather than on the actual number lost due to disaster.

So far we have presented some of the main results of Stochastic Processes and models with special reference to population point processes. Before we proceed further to develop further models relevant to problems of cavity radiation and detection, we present the physics of cavity radiation.

5. Cavity radiation

The phenomenon of cavity radiation can be viewed fundamentally as a consequence of the resonant interaction of the electromagnetic field with a microwave cavity endowed with various currents. The currents can be caused by an atomic beam passing the cavity, in which case the interaction is essentially by the coupling of the electric dipole moments to the field. Alternatively we may have a solid body containing magnetic dipole moments; there are other possibilities too, like, for example, currents (flowing) in the walls causing dissipation of the field. Thus, generally speaking we have the electromagnetic field coupled to the material substance (called cavity) carrying the charge or the dipole moment. Our emphasis will be on the physical process by which energy in the material system may be systematically transferred to the electromagnetic field; thus the field may be strongly amplified or forced to go into a steady state of oscillation. In this process the field may become very strong, leading to almost a classical situation. Since the amplification of the field can be detected only by measurements, the detection process also plays a vital role; thus cavity radiation and detection go together.

Since the amplification is essentially by stimulated action, the terminology laser, standing for **L**ight **A**mplification by **S**timulated **E**mission of **R**adiation, has come to stay. There are many approaches to the problem; there is a long history starting from the discovery of photoelectric effect by Einstein. Although the photoelectric effect provided great stimulus for the development of quantum mechanics, there has been considerable lag in the use of a fully quantized theory to problems involving radiation and its detection. We shall not discuss here the merits and demerits of semi-classical approaches; it is sufficient at this point to note that photon statistics which is characteristic of radiation has to be treated correctly, and that wave–particle duality which is essential to any treatment will be lost in transit if we resort to a semi-classical approach. Moreover there are also other non-classical features of light like anti-bunching and squeezing which cannot be brought within the ambit of a semi-classical theory. In this chapter we shall present the basic results that follow from a fully quantized theory. We first present the density matrix treatment of a cavity–field system. We then establish the Markov property and show how the Markov property renders the analysis simple and straightforward. We then deal with the detection process from many angles. The final section is devoted to a short discussion on special correlation functions known as coherence functions. The quantum process of detection is then described through the coherence functions, and the product densities of the emission process are identified. The reader is assumed to be familiar with quantum mechanics and its applications to problems involving radiation fields.

5.1   Quantum approach to cavity radiation: density matrix

We now proceed to discuss the fully quantum mechanical theory of a laser. For the sake of clarity we confine our attention to a single-cavity mode. To describe laser oscillation, we must consider essentially a non-linear active medium together with a damping mechanism. To obtain amplification, we introduce two-level atoms in the upper state denoted by $|a\rangle$ at random times. We include the dissipative action in the theory by coupling the electromagnetic field to rapidly decaying and therefore non-resonant atoms injected into the cavity in the lower state $|b\rangle$. The states $|a\rangle$ and $|b\rangle$ of the atom are assumed to decay as in the Wigner–Weisskopf theory of radiation damping. Let $|c\rangle$ be the state into which the atom in the state $|a\rangle$ decays with emission of non-resonant radiation with a decay constant γ_a. Likewise let $|d\rangle$ be the state to which the atom in the state $|b\rangle$ decays with a decay constant γ_b. First we consider the change in the density matrix due to a single atom being injected into the upper state $|a\rangle$ at time t_0; using the number representation, the change over a duration T (which is long compared to the atomic lifetime, but short compared to the characteristic time of growth/decay of the laser radiation) is given by

$$5.1.1 \qquad \delta\rho_{nn'} = \rho_{nn'}(t_0 + T) - \rho_{nn'}(t_0).$$

The change in the density matrix is due to the time-evolution of the combined atom–laser field system over the interval $(t_0, t_0 + T)$ and hence we must resort to a more detailed description of the density matrix element, taking into account the atomic states as well. Finally we can trace over the atomic states; thus we have

$$5.1.2 \qquad \rho_{nn'}(t_0 + T) = \sum_{\alpha} \rho_{\alpha n, \alpha n'}(t_0 + T),$$

where α takes on the values a, b, c, d. We note that if $\hbar H$ is the Hamiltonian describing the interaction of the active atom with the single-mode radiation field, then we have

5.1.3 $\qquad H = H_{rad} + H_{atom} + V = H_0 + V$

$\qquad\qquad H_{rad} = v a^+ a, \quad H_{atom} = W_a \sigma^+ \sigma + W_b \sigma \sigma^+,$

5.1.4 $\qquad V = g(a^+ \sigma + a \sigma^+),$

where v is the laser frequency to be determined from the theory (as eigen-value) and $\hbar W_a$ and $\hbar W_b$ are the atomic energies; the operators σ and σ^+ are the lowering and raising operators operating on the atomic states

5.1.5 $\qquad |a\rangle = \begin{pmatrix} 1 \\ 0 \end{pmatrix}, \quad |b\rangle = \begin{pmatrix} 0 \\ 1 \end{pmatrix}$

and are given by

5.1.6 $\qquad \sigma = \begin{pmatrix} 0 & 0 \\ 1 & 0 \end{pmatrix}, \quad \sigma^+ = \begin{pmatrix} 0 & 1 \\ 0 & 0 \end{pmatrix}.$

The constant g is the coupling constant defined by

5.1.7 $\qquad g = e x_{ab} \varepsilon / (\sqrt{2} \hbar)$

where $e x_{ab}$ is the dipole matrix element connecting the states $|a\rangle$ and $|b\rangle$ and ε is the strength of the field.

A more formal method of attack would be to consider the states c and d as well and enlarge the Hamiltonian to include non-resonant transition to those states as well; it is, however, prudent to study these transitions in the Wigner–Weiskopf approximation. The derivation is pretty long and we give below the final set of equations that we need:

5.1.8 $\qquad \dot{\rho}_{an,an'} = -i[(H_0 + V), \rho_{an,an'}] - \gamma_a \rho_{an,an'}$

5.1.9 $\qquad \dot{\rho}_{bn+1,bn'+1} = -i[(H_0 + V), \rho_{bn+1,bn'+1}] - \gamma_b \rho_{bn+1,bn'+1}$

5.1.10 $\qquad \dot{\rho}_{an,bn'+1} = -i[(H_0 + V), \rho_{an,bn'+1}] - \gamma_{ab} \rho_{an,bn'+1}$

5.1.11 $\qquad \dot{\rho}_{bn+1,an'} = -i[(H_0 + V), \rho_{bn+1,an'}] - \gamma_{ab} \rho_{bn+1,an'}$

5.1.12 $\qquad \dot{\rho}_{cn,cn'} = \gamma_a \rho_{an,an'}, \, \dot{\rho}_{dn+1,dn'+1} = \gamma_b \rho_{bn+1,bn'+1}$

where γ_{ab} is given by

5.1.13 $\qquad \gamma_{ab} = \tfrac{1}{2}(\gamma_a + \gamma_b).$

The commutators on the r.h.s. can be expanded; for instance, 5.1.8 can be rewritten in the form

$$\dot{\rho}_{an,an'} = -i\big[(n-n')v - i\gamma_a\big]\rho_{an,an'}$$

$$5.1.14 \qquad -i\big[V_{an,bn+1}\rho_{bn+1,an'} - \rho_{an,bn'+1}V_{bn'+1,an'}\big].$$

In a similar way equations 5.1.9 through 5.1.11 can be rewritten as

$$\dot{\rho}_{bn+1,bn'+1} = -i\big[(n-n')v - i\gamma_b\big]\rho_{bn+1,bn'+1}$$

$$5.1.15 \qquad -i\big[V_{bn+1,an}\rho_{an,bn'+1} - \rho_{bn+1,an}V_{an',bn'+1}\big]$$

$$\dot{\rho}_{an,bn'+1}$$

$$= -i\big[(n-n')v + \omega - v - i\gamma_{ab}\big]\rho_{an,bn'+1}$$

$$5.1.16 \qquad -i\big[V_{an,bn+1}\rho_{bn+1,bn'+1} + \rho_{an,an'}V_{an',bn'+1}\big]$$

$$\dot{\rho}_{bn+1,an'}$$

$$= -i\big[(n-n')v - (\omega - v) - i\gamma_{ab}\big]\rho_{bn+1,an'}$$

$$5.1.17 \qquad -i\big[V_{bn+1,an}\rho_{an,an'} - \rho_{bn+1,bn'+1}V_{bn'+1,an'}\big]$$

where ω is the atomic transition frequency.

Thus we have a closed set of linear differential equations for the elements of the density matrix; instead of solving them directly, we proceed as follows. We introduce the notation

$$\int_{t_0}^{t_0+T} \rho_{\alpha\alpha'}(t')\,dt' = \sigma_{\alpha\alpha'}(t_0,T), \quad \alpha,\alpha' = 1,2$$

$$5.1.18 \qquad \sigma_{11} = \sigma_{an,an'}, \quad \sigma_{12} = \sigma_{an,bn'+1}$$

$$\sigma_{21} = \sigma_{bn+1,an'}, \quad \sigma_{22} = \sigma_{bn+1,bn'+1}:$$

the c- and d-state matrix elements follow from 5.1.12 and the initial conditions

$$5.1.19 \qquad \rho_{cn,cn'}(t_0+T) = \gamma_a\sigma_{11}, \quad \rho_{dn+1,dn'+1}(t_0+T) = \gamma_b\sigma_{22}.$$

We next integrate both sides of 5.1.14 through 5.1.17 to obtain

$$\rho_{an,an'}(t_0+T) - \rho_{an,an'}(t_0)$$

$$5.1.20 \qquad = -\gamma_a\sigma_{11} - i\big[V_{an,bn+1}\sigma_{21} - \sigma_{12}V_{bn'+1,an'}\big]$$

$$\rho_{bn+1,bn'+1}(t_0+T) - \rho_{bn+1,bn'+1}(t_0)$$

$$5.1.21 \qquad = -\gamma_b\sigma_{22} - i\big[V_{bn+1,an}\sigma_{12} - \sigma_{21}V_{an',bn'+1}\big]$$

$$\rho_{an,bn'+1}(t_0+T) - \rho_{an,bn'+1}(t_0)$$

5.1.22
$$= -[i(\omega-v)+\gamma_{ab}]\sigma_{12} - i[V_{an,bn+1}\sigma_{22} - \sigma_{11}V_{an',bn'+1}]$$

$$\rho_{bn+1,an'}(t_0+T) - \rho_{bn+1,an'}(t_0)$$

5.1.23
$$= -[-i(\omega-v)+\gamma_{ab}]\sigma_{21} - iV_{bn+1,an}\sigma_{11} - \sigma_{22}V_{bn'+1,an'},$$

where we have replaced throughout $\rho_{nn'}$ by $\rho_{nn'}\exp+i(n-n')vt$. This of course means that in what follows ρ has to be interpreted as being represented in an interaction picture. Next we note that T is large compared with the atomic lifetime; in such a case the first term on the l.h.s. of 5.1.20 through 5.1.23 can be dispensed with. Further initial conditioning demands that the second term on the l.h.s. of 5.1.21 through 5.1.23 can be replaced by zero. Hence we have

5.1.24
$$\rho_{an,an'}(t_0) = \rho_{nn'}(t_0).$$

Thus, solving the resulting set of linear algebraic equations for $\sigma_{11}, \sigma_{12}, \sigma_{21}$ and σ_{22}, we have

5.1.25
$$\gamma_a\sigma_{11} = \rho_{nn'}(t_0)\{1 - [(n+1)R_{nn'} + (n'+1)R^*_{n'n}]\}$$

5.1.26
$$\gamma_b\sigma_{22} = \rho_{nn'}(t_0)[R_{nn'} + R^*_{n'n}](n+1)(n'+1),$$

where

$$R_{nn'} = g^2[\gamma_b(\gamma_{ab}+i\Delta) + g^2(n-n')]/$$

$$[\gamma_a\gamma_b(\gamma_{ab}^2+\Delta^2) + 2\gamma_{ab}^2 g^2(n+n'+2) + g^2(n-n')$$

5.1.27
$$\times \{g^2(n'-n) + i\Delta(\gamma_a - \gamma_b)\}]$$

where $\Delta = \omega - v$. We next note that

$$\rho_{nn'}(t_0+T) = \rho_{an,an'}(t_0+T) + \rho_{bn,bn'}(t_0+T)$$

5.1.28
$$+ \rho_{cn,cn'}(t_0+T) + \rho_{dn,dn'}(t_0+T),$$

where the first two terms are zeros since T is large compared to the atomic lifetime, and hence it follows that

5.1.29
$$\rho_{nn'}(t_0+T) = \gamma_a\sigma_{11} + \gamma_b\sigma_{22} \quad (n \to n-1, n' \to n'-1).$$

Thus we finally have, on using 5.1.25 and 5.1.26 in 5.1.1,

$$\delta\rho_{nn'} = -[(n+1)R_{nn'} + (n'+1)R^*_{n'n}]\rho_{nn'}$$

5.1.30
$$+ [R_{n-1n'-1} + R^*_{n'-1n-1}](nn')^{1/2}\rho_{n-1n'-1}.$$

The change in the density matrix estimated by the above formula is due to the injection of one atom in the upper level. If we consider a time-interval Δt which is very long compared to the atomic lifetime and small compared to the characteristic time corresponding to the (laser) radiation field, then the time-scale for description of the density matrix becomes macroscopic and the change in $\rho_{nn'}$ due to many atoms that have been injected into the system during the time-interval Δt is simply given by

$$\Delta\rho_{nn'}(t) = r_a \delta\rho_{nn'}$$

$$= -\Delta t[(n+1)R_{nn'} + (n'+1)R^*_{n'n}]\rho_{nn'}(t)r_a$$

$$5.1.31 \qquad + \Delta t[R_{n-1,n'-1} + R^*_{n'-1,n-1}](nn')^{1/2}\rho_{n-1n'-1}(t)r_a,$$

so that the coarse-grained time derivative of $\rho_{nn'}(t)$ is simply the r.h.s. of 5.1.31 divided by Δt. It is easy to see that the result of injecting upper-level atoms into the system results in some kind of a gain for the radiation field, the change in the density matrix due to the gain being specified by the coarse-grained derivative of $\rho_{nn'}(t)$.

Next we take into account the dissipation due to the non-resonant type of interaction of the cavity atoms with the field. In this case we have a sub-system consisting of non-resonant two-level atoms injected into the lower state $|b'\rangle$ of the states $|a'\rangle$ and $|b'\rangle$. The calculations run on lines parallel to those leading to 5.1.29. The corresponding σ-functions defined analogously are given by

$$5.1.32 \qquad \gamma_{a'}\sigma_{11} = 2\gamma_{a'}\gamma_{a'b'}g^2((n+1)(n'+1))^{1/2}\rho_{n+1,n'+1}(t_0)/D$$

$$\gamma_{b'}\sigma_{22} = -i\gamma_{b'}\left[i\gamma_{a'}(\Delta^2 + \gamma^2_{a'b'}) + g^2(n'+1)(\Delta + i\gamma_{a'b'})\right.$$

$$5.1.33 \qquad \left. - g^2(n+1)(\Delta - i\gamma_{a'b'})\right]\rho_{n+1n'+1}(t_0)/D,$$

where

$$D = \gamma_{a'}\gamma_{b'}(\gamma^2_{a'b'} + \Delta^2) + 2g^2\gamma^2_{a'b'}(n+1+n'+1)$$

$$5.1.34 \qquad + g^2(n'-n)[g^2(n'-n) + i\Delta(\gamma_{a'} - \gamma_{b'})].$$

The change in the density matrix can be calculated as before:

$$\delta\rho_{nn'} = \gamma_{a'}\sigma_{11} + \gamma_{b'}\sigma_{22}$$

$$5.1.35 \qquad \times (n \to n-1, n' \to n'-1) - \rho_{nn'}(t_0).$$

However, in this case we retain only the lowest-order terms since the damping constant is large; thus, if $r_{b'}$ is the rate of injection of lower-state atoms into the cavity, the loss term due to the decay of the radiation field is given by

$$\Delta\rho_{nn'}(t)\big|_{\text{loss}} = -\tfrac{1}{2}C(n+n')\rho_{nn'}(t)\Delta t$$

$$5.1.36 \qquad\qquad + C[(n+1)(n'+1)]^{1/2}\rho_{n+1\,n'+1}(t)\Delta t,$$

where C can be interpreted as the cavity bandwidth and is given by

$$5.1.37 \qquad C = v/Q = 2r_{b'}(g^2/\gamma_{b'})\gamma_{a'b'}[\gamma_{a'b'}^2 + \Delta^2]^{-1},$$

where Q can be interpreted as the cavity quality factor corresponding to the frequency.

Thus 5.1.31 and 5.1.36 give the change in the density matrix over a time-interval Δt which is large compared to the atomic lifetime and small compared to the time-interval characteristic of growth/decay of the laser field radiation. It is to be specially noted that the density matrix itself is in the interaction representation since we have absorbed the factor $\exp +i(n-n')vt$ in $\rho_{nn'}(t)$. We now proceed to form the evolution equation and discuss its general properties.

5.2 Quantum evolution equations: coarse graining and Markov property

So far we have treated in detail the mechanism of the growth of the laser field and the decay due to the loss arising from absorption by cavity atoms. There is a pumping mechanism which pumps the upper level (resonant) atoms into the cavity at rate r_a; there is also a pumping of lower level (non-resonant) atoms into the cavity at a rate $r_{b'}$. The marginal description of the laser field is obtained by summing over the atomic states of both resonant and non-resonant atoms; the gain due to resonant interaction of the atoms of the upper level a over the time-interval $(t, t + \Delta t)$ is given by 5.1.31, while the loss due to absorption of radiation by the non-resonant lower-level atoms over the interval $(t, t + \Delta t)$ is given by 5.1.36. Combining the gain and loss, we can describe the variation of the density matrix of the laser field by

$$\left(\frac{\mathrm{d}\rho_{nn'}(t)}{\mathrm{d}t}\right)_{\text{coarse}}$$

$$= -\left\{[(n+1)R_{nn'} + (n'+1)R^*_{n'n}]r_a + \frac{C}{2}(n+n')\right\}\rho_{nn'}(t)$$

$$+ (R_{n-1\,n'-1} + R^*_{n'-1\,n-1})r_a(n'n)^{1/2}\rho_{n-1\,n'-1}(t)$$

$$5.2.1 \qquad + C[(n+1)(n'+1)]^{1/2}\rho_{n+1\,n'+1}(t),$$

an equation known as the *quantum evolution equation*. It is specially to be noted that the derivative is with respect to coarse-grained time. Next the rate of change of the density matrix depends only on the number of quanta; this enables us to view the field as a population of quanta evolving in a Markov manner. There is yet another noteworthy feature; the evolution equations couple only elements of the density matrix along lines parallel to the main diagonal; for instance, the contribution to the diagonal elements arises only from the diagonal elements.

Let us consider the diagonal elements of the density matrix. We will assume that the laser field is tuned to the atomic resonant frequency ω so that $\Delta = 0$. In this case the evolution equation 5.2.1 reduces to

$$\frac{d\rho_{nn}(t)}{dt} = -\{Cn + A(n+1)/[1 + (n+1)B/A]\}\rho_{nn}(t)$$

$$+ \{An/[1 + nB/A]\}\rho_{n-1n-1}(t)$$

5.2.2
$$+ C(n+1)\rho_{n+1n+1}(t),$$

where

5.2.3 $\qquad A = 2g^2 r_a/\gamma_a\gamma_b, \quad B = 8r_a(g^2/\gamma_a\gamma_{ab})(g^2/\gamma_a\gamma_b)$

and C is the cavity bandwidth. Normally the diagonal elements of the density matrix are by definition non-negative and admit a direct probabilistic interpretation. It is not clear in our case whether the same considerations are applicable, for we have, in the process of estimation of the gain and loss terms over the interval $(t_0, t_0 + T)$, effected a coarse-grained time-average of ρ. Hence we have to make the assumption that $\rho_{nn}(.)$ are non-negative; this would of course mean that the evolution of the system over phase is so delicately balanced as to keep $\rho_{nn}(.)$ positive. Then 5.2.2 can be interpreted as the equation describing the flow of probability of finding n photons in the system. If we denote by $X(t)$ the number of photons in the cavity, then we have a population process $\{X(t)\}$ which evolves, with respect to the coarse-grain time parameter t, as a Markov chain with jumps of unity, with transition probabilities specified by

$$Pr\{X(t + \Delta t) = m \mid X(t) = n\}$$

5.2.4 $\qquad = \dfrac{A(n+1)}{1 + \dfrac{B}{A}(n+1)} \delta_{mn+1} + Cn\delta_{mn-1} \quad \Delta t + o(\Delta t) \quad m \neq n,$

$$Pr\{X(t + \Delta t) = n \mid X(t) = n\}$$

$$5.2.5 \qquad = 1 - \left\{ \frac{A(n + 1)}{1 + \dfrac{B}{A}(n + 1)} + Cn \right\} \Delta t + o(\Delta t).$$

The Markov chain is indeed t-homogeneous inasmuch as the transition probabilities do not depend on t. Thus equation 5.2.2 can be identified as the Kolmogorov forward equation governing the cavity population process $\{X(t)\}$ of photons; the birth rate and the death rate at population level n are given by

$$5.2.6 \qquad \lambda_n = A(n + 1)/[1 + B(n + 1)/A]$$

$$5.2.7 \qquad \mu_n = Cn.$$

Thus the process can be viewed as a birth-and-death process. These results hold good even if the laser field is not tuned to the atomic resonant frequency.

The solution of equation 5.2.2 can be obtained. The case of physical interest corresponds to the steady-state solution when t is large. If ρ_{nn} is the equilibrium solution, we have

$$5.2.8 \qquad \rho_{nn} = K\left(\frac{A}{C}\right)^n \Big/ \prod_{k=0}^{n}\left(1 + \left(\frac{B}{A}\right)k\right),$$

where k is a normalization constant determined by the condition $\sum_0^\infty \rho_{nn} = 1$. There are three cases of interest from the point of view of practical laser operation:

Case (i): Laser above threshold   This corresponds to the case $A > C$. In this case the probabilities ρ_{nn} start increasing from $n = 0$ up to a value of n and thereafter decrease with increasing n.

Case (ii): Laser at threshold $(A = C)$ The probability ρ_{nn} has a maximum at $n = 0$ and decreases in a roughly Gaussian fashion as n increases.

Case (iii): Laser below threshold $(A < C)$ In this case we may drop the factor $1 + Bk/A$ on the r.h.s. of 5.2.8 and write

$$5.2.9 \qquad \rho_{nn} = \left(1 - \frac{A}{C}\right)\left(\frac{A}{C}\right)^n.$$

This corresponds to the *radiation emanating from black-body cavity*; we can identify the effective temperature θ by

5.2.10 $\qquad \dfrac{A}{C} = e^{-h\nu/k_B\theta}.$

Thus for a laser below threshold, the gain term is essentially linear; in other words, the birth rate λ_n is simply equal to $(n+1)A$. This leads to the following interesting interpretation of the basic population process $\{X(t)\}$. The black-body radiation from a cavity is an assembly of photon population in which

(i) each photon, independently of other photons, has a constant probability A per unit time of emitting (creating) another photon;

(ii) photons are created at a constant rate A per unit time, this being independent of the size of the population of photons;

(iii) each photon of the population has a constant probability C per unit time of being absorbed by the cavity.

In other words, the cavity population of photons evolves as a population process with birth, death and immigration, where birth and immigration rates are equal, with the members of the population evolving independently. This is a feature characteristic of chaotic radiation, and the identification leads to considerable simplification in the analysis. If we look at the two-level atomic systems, the process of emission of radiation can be interpreted as follows: each photon present in the system *stimulates* the atom at a-level to make an emission at constant rate A per unit time; apart from this, photons are emitted spontaneously at a constant rate A per unit time. Einstein introduced these two processes of *stimulated* and *spontaneous emission* when he proposed his theory of black-body radiation. He actually assumed different rates for stimulated and spontaneous emissions and argued that the two rates should be equal if there is to be (thermal) equilibrium. The factor $(n+1)A$ representing the rate of emission arises in a natural way in the rate equation, since we have used second quantized theory which incorporates the correct photon statistics.

The off-diagonal elements $\rho_{nn'}(t)$ are also equally interesting, for these will lead to the characterization of the spectral profiles of the resulting radiation. In fact, equation 5.2.1 can be solved for $n \neq n'$ but we shall not discuss this aspect. We refer the reader to the original papers of Scully and Lamb (1967, 1968).

5.3 Detection process

Next we examine the effects due to the introduction of a detector into the system. As we have observed earlier, detection and cavity evolution go together. Nevertheless we attempt to isolate detection from evolution by assuming that a plaque of steady-state quanta (obtained after equilibrium is reached) is presented to the detector. We can assume that all the lasing and damping atoms are removed from the cavity, thus ensuring no gains or loss to the density matrix over the time-interval during which detection is performed. Then the density matrix $\rho(t)$ in the interaction picture satisfies the equation

5.3.1 $\qquad \dot{\rho}(t) = -i[V_d(t), \rho(t)],$

where $\hbar V_d(t)$ is the interaction Hamiltonian for the laser field-detector system given by

5.3.2 $\qquad \hbar V_d(t) = -e\hbar \sum_{i=1}^{N} x_i(t)E(t),$

where $x_i(t)$ is the position operator of the atomic electron of the ith atom and $E(t)$ is the quantized electric field (operator). The number N is fairly large as it corresponds to the number of atoms capable of being photo-ionized.

The problem can be analyzed using the boundary condition that at $t = 0$, $\rho(0)$ corresponds to the steady-state distribution implied by the evolution equation 5.2.1. We do not propose to present the analysis; it can be shown that if the cavity initially consists of n photons for which the probability is ρ_{nn}, then the probability that m of these photo-ionize the atoms of the detector over the interval $(0, T)$ is

5.3.3 $\qquad \pi_{m,n}(T) = \rho_{nn}\binom{n}{m}\eta^m(1-\eta)^{n-m},$

where

$$\eta = 1 - e^{-\gamma t}, \quad \gamma = 2rN,$$

where r is a constant directly proportional to the density of states. Full details of the calculation are given by Scully and Lamb (1969) and the reader may refer to the paper. The marginal distribution $q_m(T)$ of the number of photons detected or the photoelectrons emitted can be obtained by summing over n:

$$q_m(T) = \sum_{n=0}^{\infty} \rho_{nn}\binom{n}{m}\eta^m(1-\eta)^{n-m}.$$

For a full understanding of the problem, it is worth while to study photodetection taking into account the variation of the field during the time-interval T of photo counting. Since the very process of detection removes quanta, it may not be a sound proposition to delink cavity–field interaction from the detection process. After all, the field equally interacts with all the atoms whether they form part of the system meant to amplify it or dissipate it, or whether they belong to the detector system. Hence we can study the density matrix for the combined field-cavity-detector system; if $\rho_{nD,n_{D'}}(t)$ is the density matrix, the change in the density matrix can be estimated on lines exactly similar to those employed earlier for treating losses. Thus if D is the excitation factor of the detector, the change in the density matrix over the coarse-grained time-interval Δt is given by

$$(\delta\rho)_{nn}\big|_{\text{Loss due to detection}}$$

$$5.3.4 \qquad = D\Delta t\{-n\rho_{nn}(t) + (n+1)\rho_{n+1n+1}(t)\}$$

where we have summed over the states of the detector. Hence we conclude that the complete equation of motion after coarse-graining is still given by 5.2.2, provided we replace C by $C + D$. If we look at the population process $\{X(t)\}$ introduced in section 5.2, we find that the process is still Markov with the homogeneous transition probabilities defined by 5.2.4 and 5.2.5 and with C replaced by $C + D$. In particular, if we take the laser oscillation below threshold corresponding to the case $A < C + D$, we obtain the black-body cavity radiation process when it is subject to detection. Then the process $\{X(t)\}$ is a population process with individuals generating a birth-and-death process independently with
   (i)   birth rate per individual equal to A
   (ii) death rate per individual equal to C
   (iii) immigration rate equal to A
   (iv) emigration rate per individual equal to D.
The number of emigrants over an interval $(0, t)$ corresponds to the number of atoms that are photo-ionized or the number of photo-electrons emitted. If the cavity population has a steady-state distribution, then the emigration process can be interpreted as a stationary point process. In Example 3.3.3 we have established this property for the point process generated by deaths; the arguments are equally valid in this case. We shall return to this point in the next chapter when we take up a detailed study of the process.

There is an alternative way of characterizing the detection process; this uses a semi-classical analysis and describes the process of photoelectron counting in terms of instantaneous intensity of the incident light-field. The electromagnetic field is treated classically and the unperturbed electron is described quantum mechanically by a Hamiltonian H_0 endowed with a set of eigen-states corresponding to the levels E_n. It is assumed that there exist a bound state (ground state) $|\psi_\beta\rangle$ at an energy level E_β and a quasi-continuum of excited states $|\psi_k\rangle$ with energy levels $E_k = E_b + \hbar\omega_k$; the excited states have a density $\rho(\omega_k)$. The interaction between the field and the electron is switched on at time $t = 0$ and the radiation field generates a time-dependent perturbation Hamiltonian $H_I(t)$, where

$$H_I(t) = V(t) \cdot p,$$

where p is the momentum of the electron. From first-order perturbation theory, the probability of transition of the electron from the bound state to any excited state $|\psi_k\rangle$ in a time-interval Δt equals $|C_k(\Delta t)|^2$, where $C_k(t)$ is given by

5.3.5 $$C_k = \frac{1}{i\hbar} \int_0^{\Delta t} \langle\psi_k|H_I(t')|\psi_b\rangle \, e^{i\omega_k t'} \, dt'$$

The r.h.s. of 5.3.5 is easily seen to be proportional to

$$\int_t^{t+\Delta t} \langle\psi_k|P|\psi_b\rangle V(t') \, e^{i\omega_k t'} \, dt'.$$

If we further assume that the matrix element in the integrand is nearly independent of k and estimate the integral for a monochromatic radiation with frequency ω_0, then the standard analysis used in quantum mechanics leads us to conclude that the probability $\lambda(t)\Delta t$ of a transition to any of the quasi-excited states in a time-interval $\Delta t \gg 1/\omega_0$ is given by

5.3.6 $$\lambda(t) = |V|^2 \rho(\omega_0).$$

The above result is equally valid for a general radiation field not necessarily monochromatic, provided that Δt, the time-interval, is much smaller than the radiation bandwidth. The arguments we have given are based on a picture of photoemission that sounds rather over-simplified; nevertheless there is a rigorous semi-classical analysis leading to the same conclusion. The reader may refer to the paper by Mandel, Sudarshan and Wolf (1964) for a complete analysis of the problem.

Thus we have established that the probability of transition of an electron in a photodetector's surface (atoms) from its ground state to a free state during a time-interval Δt (longer than a period of electromagnetic oscillation) is proportional to the time-interval Δt and to the instantaneous intensity $T(t) = |V|^2$. For a particular realization of the instantaneous intensity, the photoelectron emissions form a Poisson (point) process with parameter $\lambda(t)$ given by 5.3.6, provided the photoemissions from different atoms are statistically independent of each other. Hence we conclude that the photoelectron emissions form a Cox process. Since in the physical problem equilibrium is established at $t=0$, the resulting process is stationary. A special case of interest arises when $V(t)$ is a stationary (complex) Gaussian process; this corresponds to the black-body cavity radiation. We have already analyzed such a situation as a mathematical problem in Examples 2.1.4 and 3.4.1. The results presented therein can be taken over *in toto* for the case when the resulting radiation is thermal in nature with a Lorentzian spectrum. It is indeed possible to deal with more general situations when the spectrum corresponds to a superposition of Lorentzian profiles; in such a case we have first to proceed as in Example 2.1.4 and then use indirect methods as in Example 3.4.1. For non-thermal radiation the method breaks down; however, the counting process is still a Cox process. The product densities of the first few orders can be readily computed using the correlational structure of the process $\{V(t)\}$.

5.4 Coherence functions and the quantum detection process

If we are interested only in the statistics of the detection process, there is an alternative method available. If $\mathbf{E}(r, t)$ is the electric field operator, it can be separated into its positive frequency part $\mathbf{E}^+(\mathbf{r}, t)$ and negative frequency part $\mathbf{E}^-(\mathbf{r}, t)$. The positive frequency part has the characteristic property

$$5.4.1 \qquad \mathbf{E}^+(\mathbf{r}, t)|\text{vacuum}\rangle = 0,$$

thus the operator $\mathbf{E}^+(\mathbf{r}, t)$ destroys a photon while its adjoint $\mathbf{E}^-(r, t)$ creates a photon. The fields $\mathbf{E}^+(r, t)$ and $\mathbf{E}^-(r, t)$ can be used to define a correlation function

$$5.4.2 \qquad G^{(1)}_{\mu\nu}(\mathbf{r}t, \mathbf{r}'t') = \text{tr}\{\rho E^-_\mu(\mathbf{r}, t)E^+_\nu(\mathbf{r}, t)\},$$

where ρ is the density-matrix operator describing the field. To simplify matters, we may only deal with only one component of the electric field. This is best done by assuming that we are interested only in photons polarized parallel to a unit complex vector $\mathbf{e}$; then the scalar-field operators E^+, E^- can be defined by

$$5.4.3 \qquad E^+(rt) = \mathbf{e}^* \cdot \mathbf{E}^+(rt)$$

$$5.4.4 \qquad E^-(rt) = \mathbf{e} \cdot \mathbf{E}^-(rt).$$

Then a hierarchy of correlation functions can be defined by

5.4.5 $\qquad G^{(1)}(\mathbf{r}t,\mathbf{r}'t') = \mathrm{tr}\{\rho E^-(\mathbf{r}t)E^+(\mathbf{r}'t')\}\, G^{(2)}(\mathbf{r}_1 t_1, \mathbf{r}_2 t_2, \mathbf{r}_3 t_3, \mathbf{r}_4 t_4)$

5.4.6 $\qquad = \mathrm{tr}\{\rho E^-(\mathbf{r}_1 t_1)E^-(\mathbf{r}_2 t_2)E^+(\mathbf{r}_3 t_3)E^+(\mathbf{r}_4 t_4)\},$

and more generally

$$G^{(n)}(\mathbf{r}_1, t_1, \mathbf{r}_2 t_2, \ldots, \mathbf{r}_{2n} t_{2n})$$
$$= \mathrm{tr}\{\rho E^-(\mathbf{r}_1 t_1)E^-(\mathbf{r}_2 t_2)\cdots E^-(\mathbf{r}_n t_n)$$
$$\times\, E^+(\mathbf{r}_{n+1} t_{n+1})\cdots E^+(\mathbf{r}_{2n} t_{2n})\}$$

To evaluate these traces we can use the coherent-state description of the electromagnetic field developed by Glauber (1963) and Sudarshan (1963). In this description of the field, we assume that the field has discrete propagation modes (corresponding to some kind of finite boundary) labelled by the index k; to describe the quantum state of the kth mode, it is convenient to introduce a set of complex amplitudes, one for each occupation state $|n_k\rangle$, $n_k = 0, 1, 2, \ldots$. The main charm of a quantum approach is its ability to characterize the phase of the different modes; in this case a large number of states $|n_k\rangle$ must be superposed. Glauber used the set of basis states of the form

5.4.7 $\qquad |\alpha_k\rangle = \sum_n [\alpha_k^n/(n!)^{1/2}]|n\rangle \exp -\tfrac{1}{2}|\alpha_k|^2,$

where α_k is an arbitrary complex amplitude; these states are called *coherent states*. We can form the direct product over all the various modes to obtain

5.4.8 $\qquad |\{\alpha_k\}\rangle = \prod_k |\alpha_k\rangle$

which are eigen-states of the destruction operator

5.4.9 $\qquad a_k|\{\alpha_k\}\rangle = \alpha_k|\{\alpha_k\}\rangle.$

The average number $\bar{n}_k$ of photons in the kth mode can be readily computed using 5.3.7:

5.4.10 $\qquad \bar{n}_k = \langle\{\alpha_k\}|a_k^+ a_k|\{\alpha_k\}\rangle = |\alpha_k|^2.$

Another interesting property is that the probability that a given coherent state has exactly n photons in a particular mode is given by

5.4.11 $\qquad P_k(n) = |\langle n_k|\{\alpha_k\}|\rangle|^2 = \dfrac{(\bar{n}_k)^n \exp(-\bar{n}_k)}{n!}$

Finally, since $E^+(\mathbf{r}t)$ is a linear combination of the annihilation operator a_k, we can identify the coherent states as the eigen-states of $E^+(\mathbf{r}t)$

$$5.4.12 \qquad E^+(\mathbf{r}t)|\{\alpha_k\}\rangle = \varepsilon(r,t)|\{\alpha_k\}\rangle,$$

where $\varepsilon(rt)$ is the c-number field obtained by replacing a_k by α_k in the expansion of $E^+(r,t)$ in terms of a complete orthonormal set of functions employing box normalization:

$$5.4.13 \qquad E^+(r,t) = i\sum (\tfrac{1}{2}\hbar\omega_k)^{1/2} u_k(\mathbf{r}) a_k \exp(-i\omega_k t),$$

where $\{u_k(r)\}$ is the complete orthonormal set of mode functions.

Now the correlation function of order n can be expressed in terms of the c-field numbers ε; we have, in fact,

$$G^{(n)}(\mathbf{r}_1 t_1, \mathbf{r}_2 t_2, \ldots, \mathbf{r}_n t_n, \ldots, \mathbf{r}_{2n} t_{2n}) = \varepsilon^*(\mathbf{r}_1 t_1, \{\alpha_k\})$$

$$5.4.14 \qquad \times \varepsilon^*(\mathbf{r}_2 t_2, \{\alpha_k\}) \cdots \varepsilon(\mathbf{r}_{n+1} t_{n+1}, \{\alpha_k\}) \cdots \varepsilon(\mathbf{r}_{2n} t_{2n}, \{\alpha_k\}),$$

a relation which demonstrates the usefulness of coherent state representation; it is to be specially noted that because of normal ordering of the operators on the r.h.s. of 5.4.6, the factorization has become possible. If we set

$$5.4.15 \qquad \mathbf{r}_{n+i} = \mathbf{r}_i, \quad t_{n+i} = t_i, \quad i = 1, 2, \ldots, n,$$

we find

$$G^{(n)}(\mathbf{r}_1 t_1, \mathbf{r}_2 t_2, \ldots \mathbf{r}_n t_n, \mathbf{r}_1 t_1, \ldots, \mathbf{r}_n t_n)$$

$$5.4.16 \qquad = \prod_{i=1}^{n} |\varepsilon(\mathbf{r}_i t_i \{\alpha_k\})|^2 \geq 0.$$

The $G^{(n)}$'s can be interpreted as product densities of the point process over space–time corresponding to the distribution of photons. These functions are known as *coincidence functions* or *inclusive probabilities*.

Next we consider the effects due to the introduction of a detector. In a fully quantized theory, the interaction Hamiltonian for the field–detector system is specified by 5.3.2. If we use first-order perturbation theory, the probability $\pi_1(t)$ of a photoemission during an observation time t is given by

$$5.4.17 \qquad \pi_1(t) = \frac{1}{2\pi} \int_0^t \int_0^t G^{(1)}(rt', rt'') S(t' - t'') \, dt' \, dt'',$$

where $S(t' - t'')$ is proportional to the square of the matrix element corresponding to the atomic transition and can be given a representation

$$5.4.18 \qquad S(t) = \int \alpha(\omega)\, e^{it\omega}\, d\omega,$$

where α can be interpreted as being the sensitivity of the detector. We will assume α to be a constant over the frequency range of the incident radiation. Then 5.4.17 reduces to

$$5.4.19 \qquad \pi_1(t) = \alpha \int_0^t G^{(1)}(\mathbf{r}t', \mathbf{r}t')\, dt'.$$

Then the counting rate $h_1(t)$ of photoemissions or the product density of degree one of electron emissions is given by

$$5.4.20 \qquad h_1(t) = \frac{d}{dt}\, \pi_1(t) = \alpha G^{(1)}(\mathbf{r}t, \mathbf{r}t).$$

The above formula can easily be generalized; the nth order product density of photoelectron emissions is given by

$$5.4.21 \qquad h_n(t_1, t_2, \ldots, t_n) = \alpha^n G^{(n)}(\mathbf{r}_1 t_1, \mathbf{r}_2 t_2, \ldots, \mathbf{r}_n t_n, \mathbf{r}_1 t_1, \mathbf{r}_2 t_2, \ldots, \mathbf{r}_n t_n).$$

It is worth noting that if we set $\alpha = 1$, the product densities describe the photon statistics of emission; in the general case, they describe the statistics of photoelectron emissions.

We henceforth drop the r-dependence since we will be concerned with the field characteristics at a fixed location. Equations 5.4.20 and 5.4.21 admit an interesting interpretation in the case of chaotic fields; for in such a case the nth order function $G^{(n)}$ can be expressed as sums of products of the first-order function $G^{(1)}$:

$$5.4.22 \qquad G^{(n)}(\mathbf{r}_1 t_1, \mathbf{r}_2 t_2, \ldots, \mathbf{r}_n t_n, \mathbf{r}_1 t_1, \mathbf{r}_2 t_2, \ldots, \mathbf{r}_n t_n) = \sum_{ij} G^{(1)}(\mathbf{r}_i t_i, \mathbf{r}_j t_j).$$

Thus the product densities correspond to a conditional Poisson point process with conditional intensity $\lambda(t)$ defined by

$$5.4.23 \qquad \lambda(t) = \alpha E^+(\mathbf{r}, t) E^-(\mathbf{r}, t).$$

If at this stage we invoke the expansion of $E^+(\mathbf{r}t)$ given by 5.4.13, and take into account 5.4.9, we can identify the α_k's as random variables having Gaussian property. Thus the intensity is the modulus square of a complex Gaussian process and we identify the process of emission of photoelectrons or the point process of detection as a doubly stochastic Poisson process with conditional intensity defined by 5.4.23. The mathematical problem discussed in Example 3.4.1 can be taken over *in toto* as the solution to the problem of determination of the generating function of the counting process of photo emissions. From this angle, we have a satisfactory solution to the problem of determination of the statistics of photoemissions corresponding to thermal radiation, with correlation function corresponding to a Lorentzian spectrum.

To sum up, cavity radiation and detection can be viewed from two angles; the first, discussed in sections 5.1 and 5.2, looks at the radiation as an assembly of photons from the second quantized field-picture, while the second identifies the process of evolution and detection in terms of Coherence functions. In the first method of description, we have been able to identify the evolution and detection of radiation as a Markov process if we are only interested in the distribution of the number of quanta. In the description in terms of coherence functions, we are able to characterize the correlation of the field intensity as well as that of photocounts directly. Questions relating to phases can be answered satisfactorily in either picture, at least to second order. However, there is an approximation that is used in either picture, namely that of coarse-graining; in the description in terms of coherence functions, this is done in a subtle way when we resort to photon counting. If we analyze the arguments that are used in passing from $\delta\rho$ to $\Delta\rho$ (see 5.1.30 and 5.1.31), the type of approximation that is used is fairly clear. A special type of coarse graining leads to Markov description; thus it is easy to see that there can be non-Markov evolution in coarse-grained time, and it is this aspect that we are going to discuss in the chapters to follow. We first make a detailed analysis of the Markov model that corresponds to the laser below threshold, and establish many interesting characteristics of black-body radiation. One interesting conclusion that emerges from this type of description is that the intensity correlations can be described in terms of the point process of photoemissions, and these also characterize to a reasonable degree the spectral properties of the resulting radiation.

Additional notes

Cavity radiation and detection had been studied from many angles. Although the process of stimulated emission is known from the time when Einstein (1917) proposed the statistical theory of black-body radiation, the idea that it could be used for amplification received importance only after the actual physical construction of the ammonia beam laser by Gordon,

Zeiger and Townes (1954). Around the same time Hanbury-Brown and Twiss (1954, 1957) measured the intensity correlations of light from a star in receivers that are separated, and showed how the measurement could be used to determine stellar diameters. This gave further impetus to the study of the properties of coherent and partially coherent beams of light; soon the process of detection assumed great importance. We have in this chapter presented only some of the salient aspects of amplification and detection, mainly based on a fully quantum mechanical approach by Scully and Lamb (1966, 1967, 1969) and that of Glauber (1963, 1969, 1970). The semi-classical approach briefly mentioned in section 5.3 is due to Mandel (1958, 1959). The reader may find it useful to consult a reprint of selected papers on coherence and fluctuating beams edited by Mandel and Wolf (1970); the collection, though not exhaustive, is fairly representative and contains a bibliography of the literature up to 1966. The book by Klauder and Sudarshan (1968) contains a very good account of coherent states and detection theory. There is an alternative approach due to Haken, who uses both semi-classical and fully quantized theory; Haken has extended the classical theory of Fokker–Planck equations to include quantum phenomena. For an interesting account of the same we refer the reader to the extended article by Haken (1970) himself in *Handbuch der Physik*. The Fokker–Planck approach has also been developed by Lax and Louisell (1967) and a detailed account is given in Louisell (1964, 1973). The recent book of Risken (1983) also deals with some of the aspects of the problem. An extensive account of photoelectron statistics is given by Saleh (1978); the bibliography contains as many as 521 references.

6. A Markov model of cavity radiation

In the previous chapter we have seen that the diagonal elements of the density matrix in the number representation satisfy a Markov evolution equation over coarse-grained time. The Markov property is a consequence of the averaging process over the time-interval large compared to the atomic lifetime and small compared to the time-scale of change of the laser field radiation. However, it is possible to accommodate non-Markov evolution in the coarse-grained time-scale, in which case the processes of emission and absorption can be viewed differently. In this chapter we consider the consequences of the Markov property for field oscillations below threshold level; in such a case we have seen that the process has a further simplicity in that the population of cavity photons evolves as a branching process, and the techniques developed in Chapter 4 are directly applicable to this situation. We first discuss the evolution and the manner in which equilibrium is reached. We then proceed to discuss the detection process. In view of the viable Markov property, we have chosen to discuss issues which are fairly difficult to deal with. For instance, the approach to equilibrium when the detector is introduced is dealt with satisfactorily. A complete analysis of the point process of emissions is then presented; an explicit expression for the correlation of photocounts in two general intervals is also provided. Problems arising from dead-time corrections to photocount distribution are also discussed, and an analysis provided leading to the determination of the statistics of dead-time corrected counts.

6.1 Cavity evolution and equilibrium distribution

Our starting-point is the quantum evolution equation 5.2.1 along with the additional loss term due to detection specified by 5.3.4. If we then look at the approximation corresponding to *oscillation below threshold*, then the cavity evolution is just the population process of the branching type dealt with in section 4.1, provided we make the following identification:

$$\lambda = v = A, \quad \mu = C, \quad \eta = D.$$

In what follows we shall use the same notation as employed in Chapter 4:

we use $X(t)$ to denote cavity photon population size at time t. The generating functions $g_1(z, t)$ and $g(z, t)$ are defined for the general case $v \neq \lambda$:

6.1.1 $\qquad g_1(z, t) = E[z^{X(t)} \,|\, X(0) = 1, v = 0]$

6.1.2 $\qquad g(z, t) = E[z^{X(t)} \,|\, X(0) = 0, v \neq 0]$

6.1.3 $\qquad g''(z, t) = E[z^{X(t)} \,|\, X(0) = n, v \neq 0],$

We take over the solutions derived in section 4.1:

6.1.4 $\qquad g_1(z, t) = 1 + (\mu + \eta - \lambda)$

$$\times (z - 1)/[\lambda(z - 1) + (\mu + \eta - \lambda z)\, e^{(\mu + \eta - \lambda)t}]$$

6.1.5 $\qquad g(z, t) = (\mu + \eta - \lambda)/[\mu + \eta - \lambda z + \lambda(z - 1)\, e^{-(\mu + \eta - \lambda)t}].$

Next we note that $g''(z, t)$ can be expressed in terms of $g_1(z, t)$ and $g(z, t)$ by using the independent nature of the population generated by individual photons:

6.1.6 $\qquad g''(z, t) = [g_1(z, t)]^n g(z, t).$

In fact, all the statistical characteristics follow essentially from $g_1(z, t)$ and $g(z, t)$. Suppose initially we have an arbitrary distribution of the population given by

6.1.7 $\qquad p_{\text{initial}}(n) = a_n \quad (n \geqslant 0)$

or equivalently in terms of the generating function:

6.1.8 $\qquad g_{\text{initial}}(z) = \phi(z) = \sum_0^\infty a_n z^n.$

Then if we denote the corresponding generating function of the cavity population at time t by $g_{\text{arb}}(z, t)$, then we have

6.1.9 $\qquad g_{\text{arb}}(z, t) = \phi(g_1(z, t)) g(z, t).$

At the outset we note that the steady state of cavity population follows from section 4.1 and is characterized by

$$g_E(z) = \lim_{t \to \infty} g_1(z, t) g(z, t)$$

6.1.10 $\qquad = \left[\dfrac{\mu + \eta - \lambda}{\mu + \eta - \lambda z}\right]^{v/\lambda}.$

In what follows we consider only the case $v = \lambda$ so that we have

6.1.11 $\qquad g_E(z) = (\mu + \eta - \lambda)/(\mu + \eta - \lambda z),$

which corresponds to the Bose–Einstein distribution characterizing thermal photons arising from black-body radiation. We wish to examine the approach to equilibrium starting from some specified initial distributions.

(i) Coherent state

If the initial state is a coherent state in the number representation, the initial distribution of cavity population is Poissonian:

6.1.12 $\qquad a_m = e^{-\bar{n}}\bar{n}^m/m!, \quad m = 0, 1, 2, \ldots .$

Thus it follows from 6.1.9 that

6.1.13 $\qquad g_{\text{coh}}(z, t) = [\exp\{[\bar{n}(z - 1)e^{-\lambda t/I}]/D(z, t)\}]/D(z, t)$

where

6.1.14 $\qquad D(z, t) = 1 + I(1 - e^{-\lambda t/I})(1 - z)$

6.1.15 $\qquad I = \lambda/(\mu + \eta - \lambda).$

The special form of dependence on z suggests that the inversion can be carried out in terms of Laguerre polynomials. Using the result

6.1.16 $\qquad [\exp -xt/(1 - t)]/(1 - t) = \sum_0^\infty L_n(x)t^n,$

where $L_n(x)$ are normalized so that $e^{-x/2}L_n(x)$ form an orthonormal family over $[0, \infty)$, we obtain

$$g_{\text{coh}}(z, t) = g_{\text{coh}}(0, t) \sum_0^\infty L_n\left(\frac{-\bar{n}\,e^{-\lambda t/I}}{I(1 - e^{-\lambda t/I})D(0, t)}\right)$$

6.1.17 $\qquad\qquad \times \left(\frac{zI(1 - e^{-\lambda t/I})}{D(0, t)}\right)^n.$

Thus we finally have

$$p_{\text{coh}}(n, t) = g_{\text{coh}}(0, t)L_n\left(\frac{-\bar{n}\,e^{-\lambda t/I}}{I(1 - e^{-\lambda t/I})D(0, t)}\right)$$

6.1.18 $\qquad\qquad \times \left(\frac{I(1 - e^{-\lambda t/I})}{D(0, t)}\right)^n,$

where $p_{\text{coh}}(n, t)$ represents the probability of finding n photons at time t when the initial cavity population is coherent. In the limit, as $t \to \infty$, $L_n(\,.\,) \to 1$, $D(0, t) \to 1 + I$, $g_{\text{coh}}(0, t) \to 1 + I$, so that the limiting distribution corresponds to the Bose–Einstein distribution. Thus if we have initially a coherent population of photons, the amplification by cavity combined with losses due to absorption and detection induces a thermal property resulting ultimately in black-body radiation.

We next investigate whether it is possible to interpret the distribution as a mixed Poisson process. The answer is in the affirmative if there exists a positive definite function $\Lambda(x, t)$ with the property

$$6.1.19 \qquad g_{\text{coh}}(z, t) = \int_0^\infty \Lambda(x, t)\, e^{-(1 - z)x}\, dx.$$

We assume that it is possible to have relation 6.1.19 satisfied, in which case

$$6.1.20 \qquad \Lambda(x, t) = \frac{1}{2\pi i} \int_{\sigma - i\infty}^{\sigma + i\infty} g_{\text{coh}}(1 - s)\, e^{-sx}\, ds.$$

Substituting 6.1.13 into the r.h.s. of 6.1.20 and evaluating the integral, we obtain

$$\Lambda(x, t) = I_0\!\left(2\sqrt{\frac{\bar{n}\, e^{-\lambda t/I}}{I(1 - e^{-\lambda t/I})} \frac{x}{I(1 - e^{-\lambda t/I})}} \right)$$

$$6.1.21 \qquad\qquad \times \exp\left[-\frac{\bar{n}\, e^{-\lambda t/I} + x}{I(1 - e^{-\lambda t/I})} \right],$$

where I_0 is the Bessel function with imaginary argument. It is easy to see that $\Lambda(x, t)$ is always positive. Thus the process is a mixed Poisson process. It is interesting to recall the identification of the photoelectron counting process as a Cox process. If the efficiency of the detector is taken to be unity, then the process describes the photon statistics. Normally this is so when we have the stationary equilibrium condition. In the present case the conditioning by restraining the initial state to a coherent one preserves the doubly stochastic Poisson character.

(ii) *Empty cavity*

When there are no photons initially, we have

$$6.1.22 \qquad g_{\text{empty}}(z, t) = g(z, t) = \sum_0^\infty \left[\frac{Iz(1 - e^{-\lambda t/I})}{D(0, t)} \right]^n \frac{1}{D(0, t)},$$

so that we have

$$6.1.23 \qquad p_{\text{empty}}(n, t) = I^n(1 - e^{-\lambda t/I})^n/[D(0, t)]^{n+1},$$

which corresponds to the Bose–Einstein distribution with expected value $I(1 - e^{-\lambda t/I})$; thus the Bose–Einstein character of the cavity population statistics is maintained throughout.

6.2 Detection process: approach to equilibrium

In this section we pose and solve an interesting problem connected with the introduction of the detector. As we observed in the introductory remarks to Chapter 5, physically amplification and detection go together since amplification can only be tested by detection. Nevertheless it is interesting to ask, at least theoretically, how the introduction of the detector changes the photon population statistics. To answer this question, we assume that the cavity population has been evolving from the distant past as a Markov population process of the branching type supported by a Poisson immigration process, so that we have an equilibrium distribution of cavity photons at the time-origin when the detector is introduced into the system. Thus the model is exactly the same as in the previous section except for the absence of the detecting mechanism over the time-interval $(-\infty, 0)$. Two questions are of interest:
 (i) the manner in which the statistical characteristics of the cavity population at $t > 0$ approach equilibrium as t is increased, and
 (ii) the statistical characteristics of the photoelectrons over an arbitrary time-interval $(t, t + T)$ and their limiting behaviour as $t \to \infty$.
Fig. 6.2.1 describes the situation in an intuitive manner. Both the questions raised above can be answered in a satisfactory manner. We

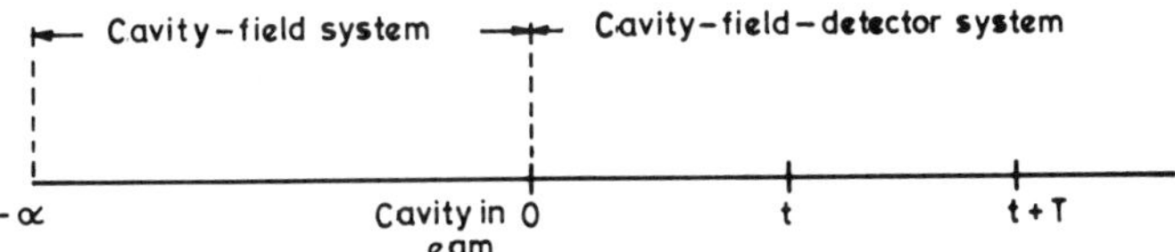

Fig. 6.2.1 Introduction of detector

follow the same notation as in section 6.1 for the generating functions of the cavity population size; in addition we introduce the following notation to characterize the photoelectron statistics:

$W(t, T)$: the counting process of photoelectrons over the interval $[t, t + T]$.

The initial distribution of the cavity population size is negative binomial, and the corresponding generating function is given by 4.1.25 with η set equal to zero. Thus we have

$$6.2.1 \qquad g_{\text{initial}}(z) = (\mu - \lambda)/(\mu - \lambda z),$$

where we have set $v = \lambda$. If we denote by $g_{CE}(z, t)$ the generating function of the population size when initially the cavity is in equilibrium with the field,

$$6.2.2 \qquad g_{CE}(z, t) = E[z^{X(t)} \mid \text{cavity in equilibrium at } t = 0,\ v \neq 0],$$

then by using the branching nature of the cavity photons, we have

$$6.2.3 \qquad g_{CE}(z, t) = g(z, t)(\mu - \lambda)/[\mu - \lambda g_1(z, t)],$$

where $g(z, t)$ and $g_1(z, t)$ are respectively given by 4.1.23 and 4.1.22. After some simple algebraic manipulations, we obtain

$$
\begin{aligned}
&g_{CE}(z, t) \\
6.2.4 \qquad &= (\mu + \eta - \lambda) \Bigg/ \left[\mu + \eta - \lambda - \lambda(z - 1)\left\{ 1 + \frac{\eta\, e^{-(\mu + \eta - \lambda)t}}{\mu - \lambda} \right\} \right],
\end{aligned}
$$

showing that the cavity population continues to possess Bose–Einstein statistics; the transient effects enter through the parameter of the distribution.

We next examine the statistics of the photoelectron population. We use the same notation as in 6.2.2 to denote the initial conditioning; thus we have

$$6.2.5 \qquad G_{CE}(z, t, T) = G(z, T)g_{CE}(G_1(z, T), t).$$

Using 6.2.4 and 4.1.24, we finally obtain

$$
\begin{aligned}
G_{CE}(z, t, T) = {}&(\xi_+ - \xi_-) \\
&\times e^{-\lambda T(1 - \xi_-)} \Bigg/ \left[(1 - \xi_-)e^{-\lambda T \Delta(z)} \right. \\
&\left. - (1 - \xi_+) - \frac{\lambda(1 - \xi_-)}{(\mu + \eta - \lambda)}(1 - \xi_+)e^{-\lambda t \Delta(z)} \right. \\
6.2.6 \qquad &\left. \times \left(1 + \frac{\eta\, e^{-(\mu + \eta - \lambda)t}}{\mu - \lambda} \right) \right].
\end{aligned}
$$

We introduce the following notation for simplicity and easy interpretation:

$$6.2.7 \qquad \bar{I} = \frac{\lambda}{\mu + \eta - \lambda}, \quad \mu + \eta - \lambda = 2\Gamma, \quad \gamma = \Gamma T,$$

$$y = \tfrac{1}{2}\lambda T \Delta = [\gamma^2 + 2\bar{I}\gamma T \eta (1 - z)]^{1/2}.$$

After some algebra, 6.2.6 can be written as

$$G_{CE}(z, t, T) = e^{y} \bigg/ \bigg[\cosh y + \frac{1}{2} \sinh y$$

$$6.2.8 \qquad \times \left\{ \left(\frac{y}{\gamma} + \frac{\gamma}{y} \right) + \frac{\eta}{(\mu - \lambda)} \left(\frac{y}{\gamma} - \frac{\gamma}{y} \right) e^{-2\Gamma t} \right\} \bigg],$$

which has a limit as t tends to infinity:

$$G_E(z, t) = \lim_{t \to \infty} G_{CE}(z, t, T)$$

$$6.2.9 \qquad = e^{y} \bigg/ \bigg[\cosh y + \frac{1}{2} \left(\frac{y}{\gamma} + \frac{\gamma}{y} \right) \sinh y \bigg].$$

The above demonstration formally proves the stationary nature of the photoelectron counting process, at least to first order. The proof can be extended to cover many time-interval counting statistics; thus we have a proof of complete stationarity of the resulting photoelectron counting process.

We recall, at this stage, our earlier identification of the photoelectron counting process from the black-body radiations as a Cox process with the conditional intensity specified by 5.4.23. The generating function of photocounts is explicitly provided in Example 3.4.1 through formula 3.4.33 which agrees with 6.2.9. Thus we have the following identification*:

$$6.2.10 \qquad \bar{I} = \frac{\lambda}{\mu + \eta - \lambda} \longmapsto \text{expected value of the (stationary) intensity of the}$$

$$\text{resulting radiation}$$

$$6.2.11 \qquad \Gamma = \frac{\mu + \eta - \lambda}{2} \longmapsto \text{width of the Lorentzian spectrum.}$$

In other words, *the Markov branching model of cavity evolution represents*

---

* We have set α, the detector efficiency occurring in 5.4.23, equal to unity. Otherwise we have to multiply the l.h.s. by a factor α. This also implies that the factor $\bar{I}\eta$ occurring in the definition of y must be replaced by $\bar{I}$.

the black-body radiation with a Lorentzian profile. Mathematically we have an interesting result:

The stationary point process of emigrations generated by a population process with constant rates λ, μ, v and η respectively of birth, death, immigration and emigration is identical with a Cox process whose intensity function is the modulus square of a zero-mean complex Gaussian Markov process, provided the birth rate λ equals the immigration rate v and the parameter Γ characterizing the correlation of the Gaussian process is chosen to be $(\mu + \eta - \lambda)/2$.

An expression for the generating function of the joint distribution of photoelectrons in two time-intervals can be derived using Markov and the branching nature of the evolution process. We introduce the twofold generating function by

$$G_E^{(2)}(z_1, t_1, T_1; z_2, t, T_2)$$

6.2.12
$$= E[z_1^{w(t_1, T_1)} z_2^{w(t_2, T_2)} \,|\, \text{cavity-field-detector system in equilibrium initially}].$$

Since the cavity-field-detector system is in equilibrium initially, it is so at the epoch t_1 as well, and hence we can set $t_1 = 0$ without loss of generality. If $t > T_1$, then there is no overlap of the intervals under consideration. For simplicity we deal with this situation. We notice that the counting process is by itself non-Markov. However, if we supplement the process by the population size process $\{X(t)\}$, then the enlarged process is Markov. Let us assume that the population size $X(t)$ takes the values n_0, n_1, n_2 respectively at the time-epochs 0, T_1 and t. Fig. 6.2.2 shows the configuration of the intervals over which photoelectron counting is performed. Our object is to

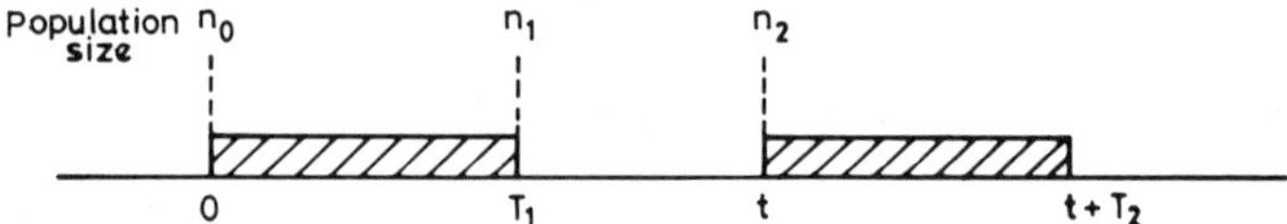

Fig. 6.2.2 Configuration of the intervals

obtain an explicit expression for $G_E^{(2)}$ as defined by 6.2.12. It is easy to make the computation block by block. Given that $X(t) = n_2$, the contribution to counts over the interval is given by

6.2.13
$$E[z_2^{W(t, t + T_2)} \,|\, X(t) = n_2] = G(z_2, T_2)[G_1(z_2, T_2)]^{n_2}.$$

Next we proceed backwards up to the epoch T_1; we have

$$E\left[z_2^{W(t,t+T_2)} \mid X(T_1) = n_1\right]$$

$$= G(z_2, T_2)g(G_1(z_2, T_2), t - T_1)[g_1(G_1(z_2, T_2), t - T_1)]^{n_1}.$$

Proceeding up to the epoch 0, we have

6.2.14
$$E\left[z_2^{W(t,t+T_2)} z_1^{W(0,T_1)} \mid X(0) = n_0\right]$$

$$= G(z_2, T_2)g(G_1(z_2, T_2), t - T_1)H(z_1, g_1(G_1(z_2, T_2), t - T_1), T_1)$$

$$\times [H_1(z_1, g_1(G_1(z_2, T_2), t - T_1), T_1]^{n_0},$$

where H and H_1 are given by 4.1.14 and 4.1.18 respectively. Finally, observing that the cavity population initially has an equilibrium distribution specified by 6.1.11, we obtain

$$G_E^{(2)}(z_1, T_1; z_2, t, T_2)$$

6.2.15
$$= \frac{G(z_2, T_2)g(G_1(z_2, T_2), t - T_1)H(z_1, g_1(G_1(z_2, T_2), t - T_1), T_1)}{1 + \dfrac{\lambda}{\mu + \eta - \lambda}[1 - H_1(z_1, g_1(G_1(z_2, T_2), t - T_1), T_1)]}.$$

For overlapping intervals, $G_E^{(2)}$ can be evaluated by the same method; we have more time-epochs to trace through in the backward direction; the algebra is straightforward, but the formula will become excessively complex. However, it should be appreciated that this is perhaps the simplest way to arrive at an explicit formula; the approach through the study of the Cox process is conceptually more difficult (see for example Srinivasan *et al.* (1973)).

The second-order correlational structure of the photocounts is in principle determined by 6.2.15. However, we shall see in the next section that the correlation of the photocounts can be derived with comparative ease when we resort to the product-density description of the counting process.

Finally we consider the situation when the detector is introduced into the system at the time-origin and removed at the epoch t (> 0). We then answer the question how the system relaxes to the equilibrium when initially the cavity and field are in equilibrium. Fig. 6.2.3 explains the state of the system. We denote the generating function of the cavity population at $t + T$ by $g_{CF}(z, t, T)$. We can obtain an explicit expression by going backwards, starting from $t + T$. We first note that

6.2.16
$$E\left[z^{X(t+T)} \mid X(t) = n, v \neq 0\right] = {}_D g(z, T)({}_D g_1(z, T))^n,$$

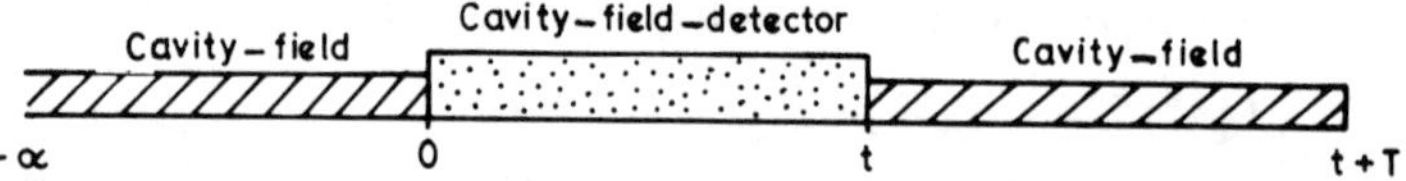

Fig. 6.2.3 Relaxation of the cavity–field system when detector is in the system over 0, t

where $_Dg$ and $_Dg_1$ are the generating functions of the detection-free population size and can be obtained from 6.1.4 and 6.1.5 by setting $\eta = 0$:

6.2.17 $\quad _Dg(z, T) = (\mu - \lambda)/[(\mu - \lambda) - \lambda(z - 1)(1 - e^{-(\mu - \lambda)T})]$

6.2.18 $\quad _Dg_1(z, T) = 1 + \dfrac{(\mu - \lambda)(z - 1)e^{-(\mu - \lambda)T}}{(\mu - \lambda) - \lambda(z - 1)(1 - e^{-(\mu - \lambda)T})}.$

Again considering the time-interval $[0, t]$, we have

6.2.19 $\quad E[z^{X(t)} \mid X(0) = m, v \neq 0] = g(z, t)[g_1(z, t)]^m.$

If we now make use of the fact that the cavity is in equilibrium, we have

$$E[z^{X(t)} \mid \text{cavity in equilibrium initially}]$$

6.2.20 $\quad = g(z, t)(\mu - \lambda)/[(\mu - \lambda) + \lambda(1 - g_1(z, t))].$

Combining 6.2.16, 6.2.19 and 6.2.20, we find

6.2.21 $\quad 1/g_{CF}(z, t, T) = \left[1 - \dfrac{\lambda}{\mu - \lambda}(z - 1)(1 - e^{-(\mu - \lambda)T})\right]$

$$\times \left[1 - \dfrac{\lambda}{\mu + \eta - \lambda} \dfrac{(z - 1)e^{-(\mu - \lambda)T}}{1 - \dfrac{\lambda}{\mu - \lambda}(z - 1)(1 - e^{-(\mu - \lambda)T})}\right.$$

$$\left. \times \left(1 + \dfrac{\eta\, e^{-(\mu + \eta - \lambda)t}}{\mu - \lambda}\right)\right]$$

from which it follows that

6.2.22 $\quad g_{CF}(z, \infty, T) = \dfrac{1}{1 - \dfrac{\lambda}{\mu - \lambda}(z - 1)\left[1 - \dfrac{\eta}{\mu + \eta - \lambda}e^{-(\mu - \lambda)T}\right]}$

6.2.23 $\quad g_{CF}(z, t, \infty) = \dfrac{1}{1 - \dfrac{\lambda}{\mu - \lambda}(z - 1)}.$

The relaxation of the cavity population to Bose–Einstein behaviour is transparent from 6.2.22 and 6.2.23.

6.3 Detection process: point process of photoemissions

Next we confine our attention to the point process of photoemissions. The process is conveniently described by product densities of photoemissions appropriately conditioned at the time-origin. There are several ways of conditioning; first we look at the situation when the cavity and field are maintained in a state of equilibrium and the detector is introduced at the time-origin. This is described in Fig. 6.3.1. We introduce

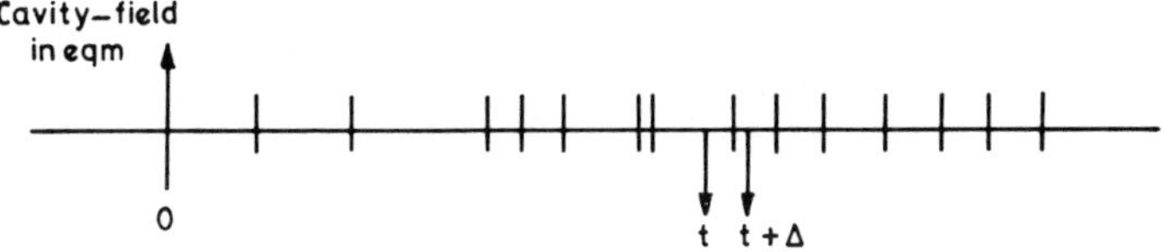

Fig. 6.3.1 Point process of emissions of photoelectrons

the product density $h_1^{CE}(t)$, where

6.3.1 $h_1^{CE}(t) = \lim_{\Delta \to 0} Pr\{W(t, \Delta) = 1 \,|\, \text{cavity field in eqm. at the origin}\}/\Delta.$

The function $h_1^{CE}(t)$ is directly proportional to the conditional expected value of $X(t)$:

$$h_1^{CE}(t) = \eta E[X(t) \,|\, \text{cavity field in eqm. at the origin}].$$

6.3.2 $$= \frac{\partial g_{CE}(z, t)}{\partial z}\bigg|_{z=1}.$$

Using the solution 6.2.4, we find

6.3.3 $$h_1^{CE}(t) = \frac{\lambda\eta}{\mu + \eta - \lambda}\left\{1 + \frac{\eta}{\mu - \lambda}e^{-(\mu + \eta - \lambda)t}\right\}.$$

If we recede the time-origin to minus infinity, we find that the product density reduces to the constant $\lambda\eta/(\mu + \eta - \lambda)$, showing that the detection process is stationary to first order. Viewed in terms of the Cox process, we note that the first-order product density is just proportional to the intensity of the resulting radiation, the constant of proportionality being the detector efficiency. Thus the first-order product density of the detection process measures the intensity of the resulting radiation, this being true for any arbitrary field.

To obtain the second-order characteristics of the point process, we introduce the immigration-free product density of emissions by

6.3.4 $\qquad {}_{I}h_1(t) = \lim_{\Delta \to 0} Pr\{W(t, \Delta) = 1 \mid X(0) = 1, v = 0\}/\Delta.$

The product density can be evaluated directly as before:

$$
{}_{I}h_1(t) = \eta E[X(t) \mid X(0) = 1, v = 0]
$$

$$
= \eta \left. \frac{\partial g_1(z, t)}{\partial z} \right|_{z-1}
$$

6.3.5 $\qquad = \eta\, e^{-(\mu + \eta - \lambda)t}.$

As before, we consider $h_2^{CE}(t_1, t_2)$, the conditional second-order product density when the origin corresponds to a point of equilibrium of cavity and field. To obtain an explicit expression for $h_2^{CE}(t_1, t_2)$, we fix our attention at the time epoch t_1 just before the emission of a photoelectron. The population tree from which a photon is detected at t_2 may have its origin at t_1, or the tree itself is formed due to spontaneous emission at an epoch later than t_1, the contribution from the two parts being mutually exclusive and exhaustive. Thus we have

$$
h_2^{CE}(t_1, t_2) = \eta E[X(t_1)(X(t_1) - 1) \mid \text{cavity-field}
$$

6.3.6 $\qquad\qquad\qquad \text{in eqm. initially}]\, {}_{I}h_1(t_2 - t_1) + h_1^{CE}(t_1)h_1^e(t_2 - t_1),$

where $h_1^e(.)$ is the product density corresponding to the initial condition when there are no cavity photons and is easily evaluated:

$$
h_1^e(t) = \eta E[X(t) \mid X(0) = 0, v \neq 0]
$$

$$
= \eta \left. \frac{\partial g(z, t)}{\partial z} \right|_{z=1}
$$

6.3.7 $\qquad = \frac{\eta \lambda}{\mu + \eta - \lambda}(1 - e^{-(\mu + \eta - \lambda)t}).$

From 6.2.4 we obtain

$$E[X(t_1)(X(t_1) - 1)\,|\,\text{cavity-field in eqm. initially}]$$

$$6.3.8 \qquad = 2\left[\frac{\lambda}{\mu + \eta - \lambda}\left(1 + \frac{\eta\,e^{-(\mu + \eta - \lambda)t}}{\mu - \lambda}\right)\right]^2.$$

Thus we finally obtain

$$h_2^{CE}(t_1, t_2) = 2\left(\frac{\eta\lambda}{\mu + \eta - \lambda}\right)^2\left[1 + \frac{\eta\{\exp(-(\mu + \eta - \lambda)t_1\}}{\mu - \lambda}\right]^2$$

$$\times \exp\{-(\mu + \eta - \lambda)(t_2 - t_1)\}$$

$$+ \left(\frac{\lambda\eta}{\mu + \eta - \lambda}\right)^2\left[1 + \frac{\eta\{\exp(-(\mu + \eta - \lambda)t_1\}}{\mu - \lambda}\right]$$

$$6.3.9 \qquad \times (1 - \exp\{-(\mu + \eta - \lambda)(t_2 - t_1)\}).$$

As before, we either recede the time-origin to $-\infty$ or take the limit as $t_1, t_2 \to \infty$ such that $t_2 - t_1$ remains finite equal to t. Denoting the limit function which is a function only of $t_2 - t_1$ by $h_{sty}(t)$, we find

$$6.3.10 \qquad h_{sty}(t) = \left(\frac{\lambda\eta}{\mu + \eta - \lambda}\right)^2(1 + e^{-(\mu + \eta - \lambda)|t|})$$

which establishes the stationarity of the point process to at least second order.

If at this stage we make use of the coherence functions introduced in section 5.4, a correspondence between the parameters can be established. For a Gaussian optical field, we have, suppressing the space dependence,

$$6.3.11 \qquad G^{(2)}(tt, t't') = |G^{(1)}(0,0)|^2 + |G^{(1)}(t, t')|^2.$$

Hence identification with the Cox process with the parameter $\alpha E^+(t)E^-(t)$ leads to

$$6.3.12 \qquad \alpha G^{(1)}(t, t')|^2 = \left(\frac{\lambda\eta}{\mu + \eta - \lambda}\right)^2 e^{-(\mu + \eta - \lambda)|t - t'|}$$

which enables us to identify $\dfrac{\mu + \eta - \lambda}{2}$ with the width of the spectrum, thus providing another confirmation of the correspondence 6.2.11.

The bunching factor $\mathscr{B}$ can be conveniently introduced in terms of the product density of degree two:

$$\mathscr{B} = h_2^{CE}(t_1, t_2)\big|_{t_2 - t_1 = 0} \big/ h_2^{CE}(t_1, t_2)\big|_{t_2 - t_2 = \infty}.$$

6.3.13
$$= 2\left\{1 + \frac{\eta}{\mu - \lambda}\exp\{-\mu + \eta - \lambda)t_1\}\right\}.$$

To start with, for small values of t_1, $\mathscr{B} > 2$; in the limit, as $t_1 \to \infty$, $\mathscr{B} = 2$. For the stationary process of photoemissions, we always have $\mathscr{B} = 2$.

The point process of photoemissions can also be characterized in terms of the intervals between successive emissions. We assume that at the time-origin we have a stationary equilibrium condition; in other words, we have initially the cavity field–detector system in equilibrium. If L_n is the nth forward recurrence time (see section 3.3), then we can provide a description of the point process of detection in terms of the family of distribution functions $F_n(.)$ (of L_n). When we have a fast detector and the average counts are far too many, it may not be prudent to describe the process in terms of $F_n(.)$; nevertheless such a description may be useful in situations where such functions can be used to estimate certain useful statistics of the process. The function $F_1(.)$ can be readily obtained by appealing to the definition:

$$\bar{F}_1(x) = 1 - F_1(x)$$
$$= Pr\{L_1 > x\}$$
$$= Pr\{W(0, x) = 0 \,|\, \text{equilibrium initially}\}$$

6.3.14
$$= G_E(0, x).$$

Using 6.2.9, we find

6.3.15
$$\bar{F}_1(x) = e^{\Gamma x} \Big/ \left[\cosh y + \frac{1}{2}\left(\frac{y}{\gamma} + \frac{\gamma}{y}\right)\sinh y\right],$$

where $y = x[\Gamma^2 + 2\bar{I}\Gamma]^{1/2}$. From 6.3.15, we can characterize the p.d.f. of the stationary interval between two successive photocounts.

To obtain $F_n(.)$ we simply note

6.3.16
$$\bar{F}_n(x) = P_E(n - 1, x),$$

where $P_E(.,x)$ is the conditional probability mass function of the photocounts over the interval $(0, x)$ when we have stationary equilibrium condition at the origin. Thus the generating function of the family of $\bar{F}_n(.)$

is given by

6.3.17 $\qquad \sum_1 \bar{F}_n(x)z^n = zG_E(z, x).$

Hence it may not be difficult to obtain explicit expressions for $\bar{F}_n(\,.\,)$, at least for the first few values of n.

We next proceed to obtain the correlation of the photocounts. First we consider non-overlapping intervals as in Fig. 6.2.2, page 96. We shall assume that the cavity population is in equilibrium at the time-origin. The correlation of the photocounts over the intervals $(0, T_1)$ and $(t, t + T_2)$ can be expressed in terms of the product densities by a straightforward adaptation of 3.3.10:

$$E[W(0, T_1)W(t, t + T_2)]$$

6.3.18 $\qquad = \int_0^{T_1} dx_1 \int_t^{T_2} h_2^E(x_1, x_2) \, dx_2,$

where h_2^E represents the product density of degree two of photocounts conditional on the time-origin being a point of stationary equilibrium. In such a case we have

$$h_2^E(x_1, x_2)$$

6.3.19 $\qquad = \left(\dfrac{\lambda\eta}{\mu + \eta - \lambda}\right)^2 (1 + \exp\{-(\mu + \eta - \lambda)(x_2 - x_1)\}).$

Thus after integration we obtain

$$E[W(0, T_1)W(t, t + T_2)]$$

$$= \left(\frac{\lambda\eta}{\mu + \eta - \lambda}\right)^2 \left[T_1 T_2 + e^{-(\mu + \eta - \lambda)(t - T_1)} \right.$$

6.3.20 $\qquad \left. \times \frac{\{1 - \exp(-\mu + \eta - \lambda)T_2\}}{\{1 - \exp(-\mu + \eta - \lambda)T_1\}}\mu + \eta - \lambda \right]$

which for equal time-intervals $(T_1 = T_2 = T)$ reduces to

$$E[W(0, T)W(t, t + T)]$$

6.3.21 $\qquad = \left(\dfrac{\lambda\eta}{\mu + \eta - \lambda}\right)^2 \left[T^2 + e^{-(\mu + \eta - \lambda)(t - T)}\left(\dfrac{1 - e^{-(\mu + \eta - \lambda)T}}{\mu + \eta - \lambda}\right) \right].$

Let us consider the rather tricky situation when the intervals have an overlap as shown in Fig. 6.3.2. For simplicity we consider only the case

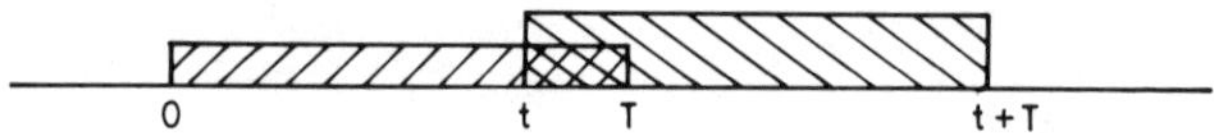

Fig. 6.3.2 Counting in overlapping intervals

$T_1 = T_2 = T$. In this case, formula 6.3.18 gets modified due to a degenerate term leading to an integral over the first-order product density over the overlapping part. Thus we have

6.3.22 $E[W(0, T)W(t, t + T)]$

$$= \int_t^T h_1^E(x)\,dx + \int_0^T dx_1 \int_t^{t+T} h_2^E(x_1, x_2)\,dx_2$$

$$= \frac{\lambda\eta}{\mu + \eta - \lambda}(T - t) + J.$$

To evaluate J we split the range of integration into several parts:

$$J = \int_0^t dx_1 \int_t^{t+T} h_2^E(x_1, x_2)\,dx_2 + \int_t^T dx_1 \int_t^T h_2^E(x_1, x_2)\,dx_2$$

6.3.23 $$+ \int_t^T dx_1 \int_T^{t+T} h_2^E(x_1, x_2)\,dx_2 = J_1 + J_2 + J_3.$$

The integrals J_1 and J_3 correspond to non-overlapping ranges for x_1 and x_2, and hence the values can be obtained by inspection from the r.h.s. of 6.3.20.

6.3.24 $$J_1 = \left(\frac{\lambda\eta}{\mu + \eta - \lambda}\right)^2\left[tT + \frac{(1 - e^{-(\mu + \eta - \lambda)T})(1 - e^{-(\mu + \eta - \lambda)t})}{(\mu + \eta - \lambda)^2}\right]$$

6.3.25 $$J_3 = \left(\frac{\lambda\eta}{\mu + \eta - \lambda}\right)^2\left[(T - t)t + \frac{(1 - e^{-(\mu + \eta - \lambda)(T - t)})(1 - e^{-(\mu + \eta - \lambda)t})}{(\mu + \eta - \lambda)^2}\right].$$

The integral J_2 is evaluated by ordering the variables x_1 and x_2:

$$J_2 = \left(\frac{\lambda\eta}{\mu + \eta - \lambda}\right)^2\left[(T - t)^2 + 2\int_0^{T-t} e^{-(\mu + \eta - \lambda)y}(T - t - y)\,dy\right]$$

$$\left[(T - t)^2 + \frac{2(T - t)}{\mu + \eta - \lambda} - \frac{2}{(\mu + \eta - \lambda)^2}(1 - e^{-(\mu + \eta - \lambda)(T - t)})\right].$$

6.3.26 $$= \left(\frac{\lambda\eta}{\mu + \eta - \lambda}\right)^2.$$

104

Combining all the terms, we finally obtain

$$E[W(0, T)W(t, t + T)]$$

$$= \frac{\lambda\eta}{\mu + \eta - \lambda}(T - t) + \left(\frac{\lambda\eta}{\mu + \eta - \lambda}\right)^2$$

$$\times \left[T^2 + \{(1 - e^{-(\mu + \eta - \lambda)(T - t)})(1 - e^{-(\mu + \eta - \lambda)t}) \right.$$

$$+ (1 - e^{-(\mu + \eta - \lambda)T})(1 - e^{-(\mu + \eta - \lambda)t})$$

$$\left. - 2(1 - e^{-(\mu + \eta - \lambda)(T - t)})\}/(\mu + \eta - \lambda)^2 + \frac{2(T - t)}{\mu + \eta - \lambda} \right].$$

6.3.27

Correlation functions corresponding to the counts in unequal and overlapping integrals can be handled in a similar way.

6.4 Dead-time corrections to the photoemissions process

We finally turn to finer aspects of the measurement of photocount statistics. The counts are usually recorded by a counter which has a *resolution time*, with the result that any photoelectron emission which is too close to the earlier one fails to get recorded by the counter. The time-interval during which recordings are not possible is known as *dead time*. Thus the recorded process of photoemission differs from the actual process; in other words, there is a *censoring* of the process. Hence it is advisable to deal with the recorded censored process so that direct contact can be established with measurements. Much of the published literature on the subject deals with the special situation when the dead time a is small, so that approximations are possible. Since we have a viable Markov model of cavity evolution, it may be worthwhile to examine the dead-time corrected emission process for arbitrary but deterministic values of a. The problem is admittedly a difficult one and we will attack it in such a way as to obtain some information on the first-order statistics of the modified process. We also provide some bounds for the second moment of the counting process.

We first concentrate on the interval between two recorded emissions. Fig. 6.4.1 shows how the emissions get censored. We have presented an exaggerated account of the detection in which 12 emissions are deleted out of a total of 19. We can visualize the process as consisting of two processes:

(i) photoemission process,
(ii) a two-valued process corresponding to the counter being free or dead.

Fig. 6.4.1 Recording of photoemissions (dead-time periods are shaded)

The two processes are interlinked. Now we look at the interval between two successive recordings when equilibrium is established. Let p_0 be the stationary probability that the counter is free at an epoch conditional on the cavity system being in equilibrium at that epoch. We choose the epoch as the origin and make an emission synchronize in an infinitesimal interval Δ around the epoch. We then analyze the course of events that contribute to the occurrence of the next recorded emission at a distance x measured from the origin. Fig. 6.4.2 illustrates the configuration. Let $q(x)$ be the

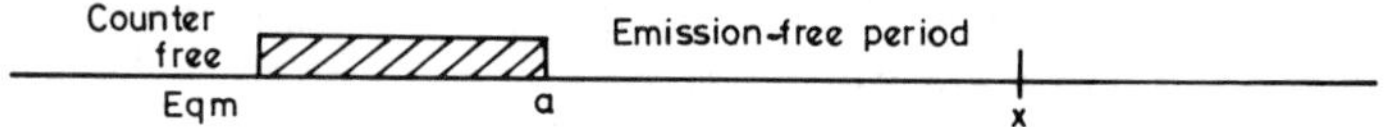

Fig. 6.4.2 Interval between two successive recordings

probability density defined by

6.4.1 $\qquad q(x) = \lim_{\Delta \to 0} Pr\{$an emission takes place in $(-\Delta, 0),$

$\qquad\qquad\qquad$ no emission in the interval $[a, x] \,|$

$\qquad\qquad\qquad$ the cavity in eqm at $-\Delta/\Delta\}.$

Then if $f_a(x)$ is the stationary probability density function governing the time-interval between two successive emissions, we have

6.4.2 $\qquad \bar{F}_a(x) = q(x) \Big/ \left[\sum_{n=0}^{\infty} p_E(n) n p_0 \eta \right],$

where $\bar{F}_a(.)$ is the survivor function corresponding to $f_a(.)$. We now proceed to calculate $q(x)$. To do this we note that if the cavity population consists of m photons at epoch a, the probability that no emission takes place is given by

6.4.3 $\qquad G(0, x - a)[G_1(0, x - a)]^m.$

Next we observe that if the cavity consists of n photons at epoch $-\Delta$ before the emission, then the probability that a recorded emission takes place in $(-\Delta, 0)$ and that the cavity consists of m photons at epoch a is given by

6.4.4 $\qquad p_E(n) n p_0 p_{\text{cav}}(m, a \mid n - 1)\eta\Delta.$

Multiplying the expressions 6.4.3 and 6.4.4 and using the definition 6.4.1, we obtain

6.4.5 $\qquad q(x) = p_0 \eta G(0, x - a) g(G_1(0, x - a), a)\phi(x - a, a),$

where $\phi(x - a, a)$ is given by

6.4.6 $\qquad \dfrac{\mathrm{d}}{\mathrm{d}z} g_E(z) \Big| z = g_1(G_1(0, x - a), a)$

and $g_E(z)$ is the generating function of the equilibrium cavity population given by

6.4.7 $\qquad g_E(z) = 1 \Big/ \left[1 + \dfrac{\lambda}{\mu + \eta - \lambda}(1 - z) \right].$

Thus we finally have

6.4.8 $\qquad \bar{F}_a(x) = \dfrac{G(0, x - a) g(G_1(0, x - a), a)}{\left\{ 1 + \dfrac{\lambda}{\mu + \eta - \lambda} [1 - g_1(G_1(0, x - a), a)] \right\}^2}$

which on further simplification reduces to

6.4.9 $\qquad \bar{F}_a(x) = \dfrac{1}{A(x)} - \dfrac{\bar{I}}{p} e^{-\Gamma(x + a)} \left\{ \dfrac{\sinh p(x - a)}{[A(x)]^2} \right\}$

with

6.4.10 $\qquad A(x) = \left[\cosh p(x - a) + \dfrac{1}{2}\left(\dfrac{p}{\Gamma} + \dfrac{\Gamma}{p} \right) \sinh p(x - a) \right] e^{-\Gamma(x - a)}$

6.4.11 $\qquad p = (\Gamma^2 + 2\Gamma\bar{I}),$

where we have used the identification 6.2.10 and 6.2.11 in terms of the average intensity and the half-width Γ. It should be noted that 6.4.2 or 6.4.9 is valid only for $x > a$; we have trivially

6.4.12 $\qquad \bar{F}_a(x) = 1 \qquad x < a.$

Now from $\bar{F}_a(x)$ we can calculate the expected value of the length of the stationary interval between two successive recordings, and from 3.3.3 it follows that this expected value equals the reciprocal of the rate of the counting process. Thus if $N_c(t)$ is the counting process representing the number of recorded counts over an arbitrary interval $(0, t)$, we have

$$6.4.13 \qquad E[N_c(t)] = rt$$

$$6.4.14 \qquad 1/r = \int_0^\infty xf(x)\,dx = \int_0^\infty \bar{F}_a(x)\,dx.$$

Using 6.4.12 and 6.4.9, we find

$$6.4.15 \qquad \frac{1}{r} = a + b\sum_0^\infty \frac{c^n}{(2n+1)p - \Gamma}\left(1 - \frac{Ib(n+1)e^{-2\Gamma a}}{(2n+3)p - \Gamma}\right)$$

where

$$6.4.16 \qquad b = 4\Gamma p/(\Gamma + p)^2, \qquad c = (p - \Gamma)^2/(p + \Gamma)^2.$$

A consistency check can be provided by setting $a = 0$. We choose η equal to unity for easy interpretation; in that case the constant p_0 occurring in 6.4.2 must be equal to unity. The verification is best done at the level of $\bar{F}_a(x)$. If we integrate 6.4.9 w.r.t. x, we obtain

$$1/r = 1/p_p\bar{I} = b\sum_0^\infty \frac{c^n}{(2n+1)p - \Gamma}$$

$$6.4.17 \qquad \times\left(1 - \frac{\bar{I}b(n+1)}{(2n+3)p - \Gamma}\right) = 1/\bar{I},$$

establishing $p_0 = 1$. We can use 6.4.17 to obtain p_0 in the general case:

$$1/p_0 = 1 + a\bar{I} + b^2\bar{I}^2$$

$$6.4.18 \qquad \times \frac{(n+1)c^n(1 - e^{-2\Gamma a})}{[(2n+1)p - \Gamma][(2n+3)p - \Gamma]}.$$

Next we consider the general case when the dead time is a random variable with the dead times arising from different recordings being independent and identically distributed with the common density function $\chi(.)$. In that case the survivor function $\bar{F}_a(x)$ as given by 6.4.9 is conditional upon the duration of the dead period being equal to a. Thus $\bar{F}(x)$, the survivor probability of the interval, is given by

$$6.4.19 \qquad \bar{F}(x) = \int_0^x \bar{F}_a(x)\chi(a)\,da + \int_x^\infty \chi(a)\,da.$$

In this case p_0 can be explicitly calculated:

$$1/p_0 = 1 + \bar{a}\bar{I} + b^2\bar{I}^2(1 - \chi^*(2\Gamma))$$

6.4.20
$$\times \sum_0^\infty \frac{c^n(n+1)}{[(2n+1)p - \Gamma][(2n+3)p - \Gamma]}$$

where $\bar{a}$ is the expected value of a and $\chi^*(.)$ the LT of $\chi(.)$.

To sum up, all that we were able to achieve in the analysis carried out so far is the average rate of counts of the recorded process and the stationary probability p_0 that the counter is open. It is desirable to have an expression for the conditional product density $h_1^c(\cdot)$; in such a case we could arrive at the second moment of the counts over an arbitrary interval. Unfortunately this is not possible and we may have to be content with the above results. However, it is possible to provide a point-wise lower bound for $h_1^c(\cdot)$.

A point-wise lower bound for $h_1^c(.)$ is the corresponding product density for the process in which the censoring is done by the type II censor. In this arrangement, each photoemission, whether recorded or not, gives rise to a dead time of fixed duration a. Fig. 6.4.3 illustrates the prolonging of the dead period by emissions that take place during the dead period; thus out of the 15 emissions only two get recorded. In this case it is easy to obtain an

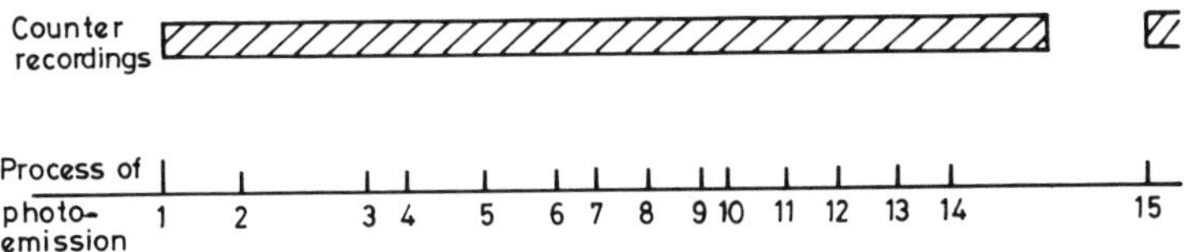

Fig. 6.4.3 Recordings of photoemissions by a Type II censor

expression for $h_1^c(.)$ by an analysis very similar to the one employed for obtaining $\bar{F}_a(.)$. First we note that for $a < x < 2a$, there is no need to resort to a bound; in fact, we have

6.4.21
$$^I h_1^c(x) = -\frac{d}{dx}\bar{F}_a(x), \qquad a < x < 2a$$

$$^I h_1^c(x) = 0, \qquad 0 < x < 1.$$

Thus it is sufficient if we obtain an expression for $^{II}h_1^c(x)$ where $x > 2a$. The events contributing to the density function are shown in Fig. 6.4.4. We can obtain the conditional product density in question by noting that if we can

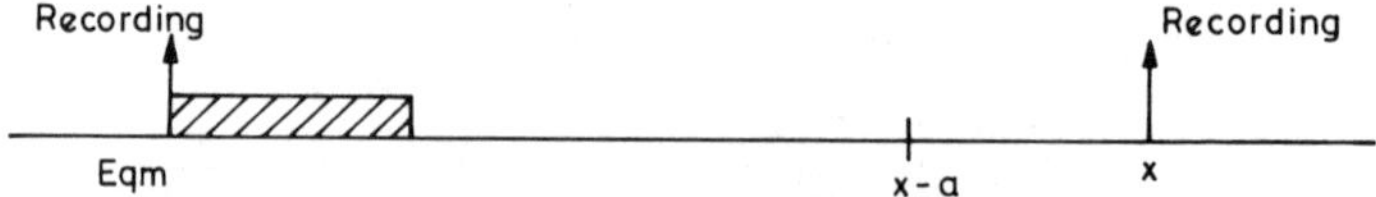

Fig. 6.4.4 Two recordings separated by an interval of length x

simply obtain the probability that no emission takes place in an interval of length b preceding x, its negative derivative w.r.t. x equals $^{II}h_1^c(x)$, provided that the value of the derivative is chosen for $b = x - a$. However, we note that the expression for the probability is simply $\bar{F}_b(x)$ defined by 6.4.2. Thus we have finally

$$6.4.22 \qquad ^{II}h_1^c(x) = -\frac{d}{dx}\bar{F}_b(x)\Big|_{b=x-a} \qquad x > 2a.$$

The conditional product density $^{I}h_1^c(.)$ satisfies the condition

$$^{I}h_1^c(x) \geqslant {}^{II}h_1^c(x) \qquad x > 2a.$$

Using the lower bound $^{II}h_1^c(x)$, we can obtain an explicit formula for the second moment of the dead-time-corrected photoelectron statistics. Using $N_c(t)$ for dead-time-corrected counts over the interval $[0, t]$, we have

$$E\{[N_c(t)]^2\} \geqslant rt + 2r \int_a^{2a} (t-x)f_a(x)\,dx$$

$$6.4.23 \qquad\qquad + 2r \int_{2a}^t f_{x-a}(x)(t-x)\,dx,$$

where

$$6.5.24 \qquad f_a(x) = -\frac{d}{dx}F_a(x).$$

For small values of a, the expression can be further simplified. The case of random dead time is more difficult to deal with and will not be attempted here.

Additional notes

The Markov model of cavity photons was first proposed by Shimoda, Takahasi and Townes (1957) who had discussed the transient as well as the steady-state characteristics of the system. Schell and Barakat (1973) fed different initial conditions to the general solution obtained by Shimoda *et al.* and discussed the approach to equilibrium. Most of the results included in sections 6.1, 6.2 and 6.3 are due to Shepherd (1981) who proposed a comprehensive model of cavity evolution and photodetection. The analysis in terms of the backward differential equation and the branching relation is based on some recent work of Srinivasan and Vasudevan (1986a, b). There is a huge literature on dead-time-corrected point processes; most of the published results relate to corrections for renewal point processes. Smith (1958) gives a short survey on counters with dead time and more generally censors. The dead-time corrections to photoelectron counting statistics for arbitrary durations of dead time were obtained by the author (1974c, 1978) using the formulation in terms of Cox processes. However, the analysis presented in section 6.4 is the result of a fruitful exchange of correspondence between the author and Dr T. J. Shepherd of RSRE during 1983.

So far we have discussed the consequences of a Markov type of evolution within the framework of branching processes. The Markov character leads to a correlation of emissions that correspond to the chaotic light (black-body radiation) with a Lorentzian spectral profile. From this angle, a natural question arises whether there can be other types of profiles. Apparently non-Markov evolution can lead to many different types of profiles, and, within the framework of population branching processes, age-dependent evolution is perhaps the simplest one can think of. Age-dependent evolution can be accommodated within the theory since the property of instantaneous emission and absorption can be thought of as having been lost in the process of coarse-graining. It is in this spirit that we proceed further to discuss the consequences of non-Markov evolution in the next two chapters.

7. Non-Markov cavity radiation – I: Bellman–Harris evolution

We now examine the consequences of non-Markov evolution. Non-Markov evolution, being residuary in our very perception, can arise in many ways. We assume that the non-Markov nature arises from some kind of age-dependence of the population growth parameters. Broadly speaking, there are two types of age-dependent population processes, namely Bellman–Harris evolution and the Kendall birth-and-death process. We deal with the former type of evolution in this chapter. A general account of such an evolution is provided in sections 7.1 and 7.2 with special reference to cavity photon population and its detection. In section 7.3 we deal with special models and recover some cases known in the literature. In the last two sections spontaneous emission is modelled in a non-Markov manner; this leads to a fairly general model which admits anti-bunching and the sub-Poissonian character of the resulting photon statistics.

7.1 Bellman–Harris evolution: general characteristics

We assume that the photons due to coarse graining evolve as a Bellman–Harris process; thus each photon, independently of other photons, has a random lifetime with probability density function $f(.)$. At the end of the life span, the photon is either replaced by two photons with probability q or absorbed by the cavity with probability $\bar{q} = 1 - q$. Spontaneous emission is incorporated in the model by assuming that there is an immigration process according to a Poisson law with parameter v. In models with constant rates, the parameter is normally so chosen that the final photon statistics has the Bose–Einstein character. Coarse graining may also result in the spontaneous emission not being at a uniform rate v; in the first instance we will keep v as a constant. Later on we will show that if spontaneous emission is inhibited, photon statistics may indicate effects like anti-bunching. In addition we assume that there is a detector present throughout and that the photo-sensitive atoms act in such a way as to absorb the photons with a uniform probability η per unit time in the coarse-grained time picture. The constancy of the expected rate of detection is the characteristic property of an ideal detector. Let $X(t)$ be the number (field intensity) of photons in the cavity and $Y(t)$ the number of photons detected over the interval $(0, t]$. It is convenient, as in section 4.1, to introduce the generating functions $g_1(z_1, z_2, t)$ and $g(z_1, z_2, t)$ by

7.1.1 $\qquad g_1(z_1, z_2, t) = \lim_{\Delta \to 0} E\big[z_1^{X(t)} z_2^{Y(t)} \,\big|\, X(0) = 1 > X(-\Delta), v = 0\big]$

7.1.2 $\qquad g(z_1, z_2, t) = E\big[z_1^{X(t)} z_2^{Y(t)} \,\big|\, X(0) = 0, v \neq 0\big].$

Next we observe that in the absence of spontaneous emission, the cavity evolution of photons is a *modified Bellman–Harris* process (the modification due to the perpetual exponential risk to which each photon is subject) and hence is a branching process with emigration. Thus the population process (of cavity photons as well as the photons removed by detection) generated by different photons are statistically independent; likewise the population processes generated by different spontaneously emitted photons are also statistically independent. Making use of the Poisson nature of the spontaneous emissions, we have

7.1.3 $\qquad \dfrac{\partial g(z_1, z_2, t)}{\partial t} = -v g(z_1, z_2, t) + v g(z_1, z_2, t) g_1(z_1, z_2, t)$

with the initial condition

7.1.4 $\qquad g(z_1, z_2, 0) = 1.$

The solution of the above equation is easily found to be

7.1.5 $\qquad g(z_1, z_2, t) = \exp\left(-v \int_0^t [1 - g_1(z_1, z_2, \tau)] \, d\tau \right).$

Thus the problem reduces to that of determining $g_1(z_1, z_2, t)$.

To obtain the equation satisfied by g_1, we note that the conditioning on the r.h.s. of 7.1.1 implies that the single initial photon is produced just prior to $t = 0$ and that the following mutually exclusive and exhaustive classes of events contribute to g_1:
(i) the initial photon survives up to t;
(ii) the photon gets absorbed by a photosensitive atom somewhere between 0 and t;
(iii) the life span of the photon comes to an end somewhere between 0 and t.

Adding these contributions, we obtain

$$g_1(z_1, z_2, t) = z_1 \bar{F}(t) e^{-\eta t} + \eta z_2 \int_0^t e^{-\eta u} \bar{F}(u) \, du$$

7.1.6
$$+ \int_0^t e^{-\eta u} f(u) \{ \bar{q} + q[g_1(z_1, z_2, t - u)]^2 \} \, du,$$

where $\bar{F}(t)$ is the complement of the distribution function corresponding to $f(t)$:

7.1.7
$$\bar{F}(t) = 1 - F(t) = \int_t^\infty f(u) \, du.$$

Equation 7.1.6 as it stands is rather intractable; however, a good deal of statistical characteristics can be obtained from it.

At the outset we assume that the function $f(.)$ and the constants η, q are such that the population process of cavity photons generated by a single photon does not explode and is ultimately extinct; it follows from the general theorem of branching processes (see Harris, 1963) that if

7.1.8
$$2q \int_0^\infty e^{-\eta t} f(t) \, dt < 1$$

then the process becomes extinct with probability one. It is to be noted that the immigration process (spontaneous emissions) stabilizes the population to a non-zero finite level with probability one. We then proceed to obtain general formulae for factorial moments. We introduce the following notation:

$$_I M_k(t) = \lim_{\Delta \to 0} E\big[(X(t)(X(t) - 1)$$

7.1.9
$$(X(t) - 2) \cdots (X(t) - k + 1) \,|\, X(0) = 1 > X(-\Delta), v = 0\big]$$

7.1.10
$$M_k = \lim_{t \to \infty} E\big[X(t)(X(t) - 1) \cdots (X(t) - k + 1)\big].$$

Next we observe that if a steady-state distribution of the population exists, it is independent of the initial size. Consequently we have

$$M_k = \lim_{t \to \infty} E\big[X(t) \cdots (X(t) - k + 1) \,|\, X(0) = 0, v \neq 0\big]$$

7.1.11
$$= \lim_{t \to \infty} \frac{\partial^k g}{\partial z_1^k} (z_1, z_2, t)\big|_{z_1 = z_2 = 1}$$

If at this stage we use the solution 7.1.5, we find

7.1.12 $$M_1 = \int_0^\infty {}_I M_1(t)\,dt$$

7.1.13 $$M_2 = \int_0^\infty {}_I M_2(t)\,dt + M_1^2.$$

The functions ${}_I M_1(t)$ and ${}_I M_2(t)$ can be obtained by differentiating both sides of 7.1.6 appropriately. Thus ${}_I M_1(t)$ and ${}_I M_2(t)$ respectively satisfy

7.1.14 $${}_I M_1(t) = \bar{F}(t)\,e^{-\eta t} + 2q \int_0^t e^{-\eta u} f(u)\,{}_I M_1(t-u)\,du$$

$${}_I M_2(t) = 2q \int_0^t e^{-\eta u} f(u)\,{}_I M_2(t-u)\,du$$

7.1.15 $$+ 2q \int_0^t e^{-\eta u} f(u)\,[{}_I M_1(t-u)]^2\,du.$$

The equations can be solved by Laplace transform technique. Denoting by ${}_I M_k^*(s)$ the LT of ${}_I M_k(t)$, we obtain

7.1.16 $${}_I M_1^*(s) = A(s)[1 - f^*(\eta + s)]/(\eta + s)$$

7.1.17 $${}_I M_2^*(s) = 2q\,A(s)\,f^*(\eta + s) \int_0^\infty e^{-st}[M_1(t)]^2\,dt,$$

where

7.1.18 $$A(s) = 1/[1 - 2q f^*(\eta + s)].$$

Thus we finally arrive at

7.1.19 $$M_1 = (v/\eta)[1 - f^*(\eta)]A^*(0)$$

and

7.1.20 $$M_2 = M_1^2 + 2vq f^*(\eta)A^*(0) \int_0^\infty [{}_I M_1(t)]^2\,dt.$$

It is indeed possible to obtain higher moments in a similar manner recursively.

The factorial moments of $Y(t)$ can be obtained using 7.1.5 and 7.1.6 and this will correspond to the moments when the origin is conditioned as in the r.h.s. of 7.1.2. However, such moments are of no interest to us; what we need are the moments of $Y(t)$ when the origin corresponds to an epoch at which the cavity photon population is in a state of equilibrium. In the case of constant birth-and-death rates, we have a simple formula expressing the generating function of equilibrium population in terms of the functions g_1 and g; however, no such formula exists for the general case. In the next section we obtain the moments by dealing with the point process of detection.

7.2 Detection process

The detection process is characterized by the counting process $\{Y(t), t > 0\}$ of photons removed by the detector over the interval $(0, t]$. As mentioned earlier, in Chapter 5, the process is also the counting process of the photoelectrons over the interval $(0, t]$ or simply the photocount process. Our main objective is to obtain the product densities of the point process when we have stationary equilibrium condition at the origin. To make progress in this direction, we first define as in Chapter 6 the conditional first-order product densities $_1h_1(t)$, $_0h_1(t)$ by

$$_1h_1(t)$$

7.2.1
$$= \lim_{\Delta,\Delta' \to 0} Pr\{Y(t+\Delta) - Y(t) = 1 \mid X(0) = 1 > X(-\Delta'), v = 0\}/\Delta$$

7.2.2
$$_0h_1(t) = \lim_{\Delta \to 0} Pr\{Y(t+\Delta) - Y(t) = 1 \mid X(0) = 0, v \neq 0\}/\Delta.$$

The functions $_1h_1$ and $_0h_1$ admit a simple interpretation: they represent the rate of production of photoelectrons (rate of detection). Second-order product densities are defined in a similar manner:

$$_1h_1(t_1, t_2) = \lim_{\Delta,\Delta',\Delta'' \to 0} Pr\{Y(t_1 + \Delta) - Y(t_1) = 1,$$

7.2.3
$$Y(t_2 + \Delta') - Y(t_2) = 1 \mid X(0) = 1 > X(-\Delta'')\}/\Delta\Delta'$$

$$_0h_2(t_1, t_2) = \lim_{\Delta,\Delta' \to 0} Pr\{Y(t_1 + \Delta) - Y(t_1) = 1,$$

7.2.4
$$Y(t_2 + \Delta') - Y(t_2) = 1 \mid X(0) = 0, v \neq 0\}/\Delta\Delta'.$$

However, we are interested in the case when the stationary equilibrium condition prevails at the origin. If we denote the corresponding conditioning by the superscript E, we note

7.2.5
$$h_1^E(t) = \text{a constant} = h_1^E$$

since the stationary equilibrium condition at the time-origin also implies the same at $t > 0$. We also have by the same token

7.2.6 $\qquad h_1^E = \lim_{t \to \infty} {}_0 h_1(t)$

since we can assume the origin to recede to $-\infty$. Next we note that for $t_2 > t_1$ the second-order product density $h_2^E(t_1, t_2)$ is only a function of $t_2 - t_1$. In fact we have

7.2.7 $\qquad h_2^E(t_1, t_2) = h_{sty}(t) = \lim_{\substack{t_1, t_2 \to \infty \\ t_2 - t_1 = t}} {}_0 h_2(t_1, t_2).$

We now proceed to obtain the equations satisfied by the product densities ${}_1 h_1(.)$ and ${}_1 h_2(...)$. To obtain contributions to ${}_1 h_1(t)$ for $t > 0$, we pursue the life history of the photon that is present (rather produced at the origin). The photon may either have its life cycle terminated between 0 and t; if it does, say between τ and $\tau + d\tau$, then with probability q it is replaced by two photons at that epoch which in turn become independent primary photons each contributing ${}_1 h_1(t - \tau)$. If, on the other hand, the photon life extends up to t, then it is photoabsorbed in the infinitesimal interval following t. Thus we have

7.2.8 $\qquad {}_1 h_1(t) = \bar{F}(t) e^{-nt} + 2q \int_0^t e^{-n\tau} f(\tau) {}_1 h_1(t - \tau) \, d\tau.$

We repeat the same arguments combined with combinatorial deduction to obtain

$$
{}_1 h_2(t_1, t_2) = 2q \int_0^{\min(t_1, t_2)} e^{-n\tau} f(\tau)
$$

7.2.9 $\qquad\qquad\qquad \times \left[{}_1 h_2(t_1 - \tau, t_2 - \tau) + {}_1 h_1(t_1 - \tau) {}_1 h_1(t_2 - \tau) \right] d\tau.$

To obtain the product density ${}_0 h_1(t)$ we note that a primary photon must be produced by spontaneous emission subsequent to the origin. Noting that the rate of emission is uniform, we obtain

7.2.10 $\qquad {}_0 h_1(t) = v \int_0^t {}_1 h_1(\tau) \, d\tau.$

The second-order product density is obtained by noting that the two photo-absorptions can be caused by a single spontaneous emission or two different emissions. Taking into account both the possibilities, we have

7.2.11 $\qquad {}_0 h_2(t_1, t_2) = {}_0 h_1(t_1) {}_0 h_1(t_2) + v \int_0^{\min(t_1, t_2)} {}_1 h_2(t_1 - \tau, t_2 - \tau) \, d\tau.$

Taking LT of both sides of 7.2.8 through 7.2.11, we have

7.2.12 $\qquad {}_Ih_1^*(s) = \eta[1 - f^*(\eta + s)]A(s)/(\eta + s)$

7.2.13 $\qquad {}_Ih_2^*(s_1, s_2) = 2q_I h_1^*(s_1)_I h_1^*(s_2)$

$$\times f^*(s_1 + s_2 + \eta)A(s_1 + s_2)$$

7.2.14 $\qquad {}_0h_1^*(s) = v_I h_1^*(s)/s$

$$_0h_2^*(s_1, s_2) = {}_0h_1^*(s_1)_0 h_1^*(s_2)$$

7.2.15 $\qquad\qquad\qquad + v_I h_2^*(s_1, s_2)/(s_1 + s_2).$

Next we use Tauberian theorem (see Titchmarch (1937) to evaluate the limit on the r.h.s. of 7.2.6. Using 7.2.12 and 7.2.14, we obtain

$$h_1^E = \lim_{s \to 0} {}_0h_1^*(s)s$$

7.2.16 $\qquad = (1 - f^*(\eta))A(0).$

The r.h.s. of 7.2.16 can be identified as ηM_1; this can be established directly by observing that the expected rate of detection is η times the expected value of the photon number (intensity). Using 7.2.7 and proceeding in a similar way, we obtain after some calculations,

7.2.17 $\qquad h_{sty}(t) = (h_1^E)^2 + 2vqf^*(\eta)A(0)J(t)$

where $J(t)$ is given by

7.2.18 $\qquad J(t) = \dfrac{1}{2\pi i} \displaystyle\int_{-i\infty}^{+i\infty} {}_Ih^{1*}(s)_I h_1^*(-s)\, e^{s|t|}\, ds.$

We can easily derive a formula for $\mathcal{B}$, the bunching factor:

7.2.19 $\qquad \mathcal{B} = \dfrac{h_{sty}(0)}{h_{sty}(\infty)} = 1 + \dfrac{2qf^*(\eta)J(0)}{v[1 - f^*(\eta)]^2 A(0)}.$

Since the second term on the r.h.s. of 7.2.19 is positive, we have established bunching for the general model in question.

7.3 Special models

We next consider some special models by making some specific choices.

7.3a Shepherd model

If we choose $f(.)$ by

7.3.1
$$f(t) = e^{-(\mu+\lambda)t}(\mu+\lambda) \qquad t \geqslant 0$$
$$= 0 \qquad\qquad\qquad t < 0$$

and the constant q by

7.3.2 $\qquad q = \lambda/(\mu+\lambda),$

then equation 7.1.6 can be converted into a differential equation by differentiating both sides w.r.t t. Thus we have

$$\frac{\partial g_1(z_1, z_2, t)}{\partial t} = -(\mu+\lambda+\eta)g_1(z_1, z_2, t)$$

7.3.3
$$+ \mu + \eta z_2 + \lambda[g_1(z_1, z_2, t)]^2$$

with the initial condition

7.3.4 $\qquad g_1(z_1, z_2, 0) = z_1.$

Equation 7.3.3 is identical with 4.1.16.

The product densities can be calculated explicitly. For we have

7.3.5 $\qquad f^*(s) = \dfrac{\mu+\lambda}{\mu+\lambda+s}, \qquad A(s) = \dfrac{\lambda+\mu+\eta+s}{\mu+\eta+s-\lambda}.$

Substituting in 7.2.16 and 7.2.18, we find

7.3.6 $\qquad h_1^E = v\eta/(\mu+\eta-\lambda)$

7.3.7 $\qquad h_{sty}(t) = \left(\dfrac{\eta}{\mu+\eta-\lambda}\right)^2 [\lambda^2 e^{-(\mu+\eta-\lambda)|t|} + v^2].$

Choosing $v = \lambda$, we recover all the results of section 6.3.

7.3b Two-stage cavity evolution model

We next choose $f(.)$ to correspond to a general Erlangian distribution:

7.3.8 $\qquad f(t) = \dfrac{\alpha\beta}{\alpha-\beta}[e^{-\beta u} - e^{-\alpha u}] \qquad (\alpha, \beta > 0).$

This distribution is used extensively in queues and telephone network theory (see Cox, 1962) and can be interpreted as the convolution of two exponential distributions with different parameters $\alpha, \beta > 0$. A special case of this arises if we take the limit $\beta \to \alpha$:

7.3.9 $\qquad f(t) = \alpha^2 t\, e^{-\alpha t},$

which is a special case of the choice 4.2.3 discussed earlier. If we take the limit $\alpha \to \infty$ in 7.3.8, we recover the Shepherd model, provided we set

7.3.10 $\qquad \beta = \mu + \lambda, \quad q = \lambda/(\mu + \lambda).$

Now we discuss the general case. The convolutory nature of the distribution has a simple interpretation; the non-Markov evolution arising from coarse-graining implies that the photon has to pass through an intermediate phase before cavity absorption or stimulated emission can take place. As in section 4.3, we split the process $\{X(t)\}$ as the sum of two processes $\{X_1(t) : t > 0\}$ and $\{X_2(t) : t > 0\}$, where $X_1(t)$ and $X_2(t)$ represent the population size respectively in phases 1 and 2. We further assume that the life span of phase 1 (2) is exponentially distributed with parameter α (β). Thus it is advantageous to study the vector process $\{X_1(t),\, X_2(t),\, Y(t) : t > 0\}$. Accordingly we introduce the generating functions $g_i(z_1, z_2, z, t)$ and $g(z_1, z_2, z, t)$, where

$$g_i(z_1, z_2, z, t)$$

7.3.11 $\qquad = E\big[z_1^{X_1(t)} z_z^{X_2(t)} z^{Y(t)} \,\big|\, X_i(0) = X(0) = 1, v = 0\big] \quad i = 1, 2$

7.3.12 $\qquad g(z_1, z_2, z, t) = E\big[z_1^{X_1(t)} z_2^{X_2(t)} z^{Y(t)} \,\big|\, X(0) = 0, v \neq 0\big].$

The vector process $\{X_1(t), X_2(t), Y(t) : t > 0\}$ is a Markov process; in view of the branching nature of the process, a number of interesting connecting relations follow. For instance, $g(z_1, z_2, z, t)$ satisfies

$$\frac{\partial g(z_1, z_2, z, t)}{\partial t} = -vg(z_1, z_2, z, t)$$

7.3.13 $$+ vg(z_1, z_2, z, t)g_1(z_1, z_2, z, t)$$

from which we obtain

7.3.14 $\qquad g(z_1, z_2, z, t) = \exp -v \int_0^t \big[1 - g_1(z_1, z_2, z, \tau)\big]\, \mathrm{d}\tau.$

If $g_{ij}(z_1, z_2, z, t)$ is defined by

$$g_{ij}(z_1, z_2, z, t)$$

7.3.15 $\qquad = E\big[z_1^{X_1(t)} z_2^{X_2(t)} z^{Y(t)} \,\big|\, X_1(0) = i, X_2(0) = j\big] \qquad i, j = 1, 2, 3, \ldots$

then the branching nature of the population process implies that

$$g_{ij}(z_1, z_2, z, t) = [g_1(z_1, z_2, z, t)]^i$$

7.3.16
$$\times [g_2(z_1, z_2, z, t)]^j g(z_1, z_2, z, t).$$

From the above relation we infer that if $g_E(z_1, z_2, z, t)$ is the generating function when the cavity is initially maintained in a state of stationary equilibrium, then $g_E(z_1, z_2, z, t)$ is given by

$$g_E(z_1, z_2, z, t) = g(z_1, z_2, z, t)$$

7.3.17
$$\times g(g_1(z_1, z_2, z, t), g_2(z_1, z_2, z, t), 1, \infty).$$

Thus it is sufficient if g_1 and g_2 are determined; if moments are desired, then the moments of g_1 and g_2 will determine the moments of the other functions.

Proceeding as in section 4.3, we obtain the backward differential equations

7.3.18
$$\frac{\partial g_1(z_1, z_2, z, t)}{\partial t} = (\alpha + \eta)g_1(z_1, z_2, z, t)$$

$$+ \alpha g_2(z_1, z_2, z, t) + \eta z$$

7.3.19
$$\frac{\partial g_2(z_1, z_2, z, t)}{\partial t} = -(\beta + \eta)g_2(z_1, z_2, z, t) + \beta \bar{q} + \eta z$$

$$+ \beta q[g_1(z_1, z_2, z, t)]^2$$

with the initial condition

7.3.20
$$g_i(z_1, z_2, z, 0) = z_i.$$

Although equations 7.3.18 and 7.3.19 cannot be solved explicitly, the moments can be generated by differentiating w.r.t. z_i or z as appropriate.

To obtain the moments, we can directly substitute in the general formulae provided in sections 7.1 and 7.2. After some computations, we obtain from 7.1.16 and 7.1.19

$$M_1(t) = \{[\alpha + \beta + \eta - \Gamma_2]\,e^{-\Gamma_2 t}$$

7.3.21
$$+ [\Gamma_1 - (\alpha + \beta + \eta)]\,e^{-\Gamma_1 t}\}/(\Gamma_1 - \Gamma_2)$$

7.3.22
$$M_1 = E[X(\infty)] = v(\eta + \alpha + \beta)/[(\alpha + \eta)(\beta + \eta) - 2q\alpha\beta],$$

where $\Gamma_{1,2}$ are given by

7.3.23
$$2\Gamma_{1,2} = \alpha + \beta + 2\eta \pm [(\alpha - \beta)^2 + 8q\alpha\beta]^{1/2}.$$

Substituting 7.3.21 in 7.1.20 and using the identification 7.3.10, we obtain

$$M_2 = M_1^2\left[1 + \frac{\lambda\alpha}{(\alpha + \mu + \lambda + 2\eta)v}\right.$$

7.3.24
$$\left. + \frac{\alpha\lambda M_1}{(\alpha + \mu + \lambda + 2\eta)(\alpha + \mu + \lambda + \eta)}\right].$$

We note that if $\alpha \to \infty$, we have the familiar result $M_2 = 2M_1^2$, provided we set $\lambda = v$; on the other hand, if α is small, bunching is reduced.

The stationary product-densities of photoemissions can be obtained from the general formulae of section 7.2:

7.3.25
$$h_1^E = \eta M_1 = \eta v(\eta + \alpha + \beta)/[(\alpha + \eta)(\beta + \eta) - 2q\alpha\beta]$$

7.3.26
$$h_{sty}(t) = \eta^2 M_1^2(1 + R(t)),$$

where $R(t)$ is given by

$$R(t) = v\beta\alpha q\{[\Gamma_1^2 - (\alpha + \beta + \eta)^2]\Gamma_2\, e^{-\Gamma_1 t}$$

$$+ [(\alpha + \beta + \eta)^2 - \Gamma_2^2]\Gamma_1\, e^{-\Gamma_2 t}\}/$$

7.3.27
$$\{M_1^2[(\alpha + \eta)(\beta + \eta) - 2q\alpha\beta]\Gamma_1\Gamma_2(\Gamma_1^2 - \Gamma_2^2)\}.$$

Next we choose $q = \frac{1}{2}$. Then we find $\Gamma_1 = \alpha + \beta + \eta$, $\Gamma_2 = \eta$. The product densities are given by

7.3.28
$$h_1^E = v$$

7.3.29
$$h_{sty}(t) = v^2[1 + R(t)]$$

where $R(t)$ now reduces to

7.3.30
$$R(t) = \frac{\alpha\beta}{2v(\alpha + \beta + \eta)}\, e^{-\eta t}.$$

If we choose α, β so that

7.3.31
$$\alpha\beta = 2v(\alpha + \beta + \eta)$$

then the radiation *corresponds to a thermal stream with a Lorentzian profile, at least to second order.*

7.3c Coherent model

This model is obtained by choosing $q = 0$. In this case we have

$$g_1(z_1, z_2, t) = z_1 \bar{F}(t)\,e^{-\eta t} + \eta z_2 \int_0^t e^{-\eta u} \bar{F}(u)\,du$$

7.3.32
$$+ \int_0^t e^{-\eta u} f(u)\,du.$$

The equilibrium distribution of the cavity photons is given by

7.3.33 $\qquad g(z_1, 1, \infty) = \exp[v(z_1 - 1)(1 - f^*(\eta))/\eta]$

a result which establishes the Poisson nature, as expected on intuitive considerations.

The steady-state rate of detection (production of photoelectrons) follows from the general formula 7.2.16:

7.3.34 $\qquad h_1^E = v(1 - f^*(\eta)).$

The higher-order product densities can be obtained directly. We note that

7.3.35 $\qquad {}_1 h_m(t_1, t_2, \ldots, t_m) = 0 \quad (m > 1).$

Hence we have the result

7.3.36 $\qquad {}_0 h_m(t_1, t_2, \ldots, t_m) = {}_0 h_1(t_1)\,{}_0 h_1(t_2) \cdots {}_0 h_1(t_m),$

where

7.3.37 $\qquad {}_0 h_1(t) = v\eta \int_0^t M_1(u)\,du$

7.3.38 $\qquad M_1^*(s) = [1 - f^*(\eta + s)]/(\eta + s).$

Thus the process $Y(t)$ is Poisson with parameter $v(1 - f^*(\eta))$; the non-Markov interaction renormalizes the parameter of the Poisson process corresponding to the coherent case.

7.4 Inhibited immigration

So far we have dealt with the models of modified Bellman–Harris processes supported by an immigration process. Since the evolution equation is described with respect to coarse-grained time, it is natural to expect the birth-and-death rates to exhibit dependence on memory, and one easy way to characterize the memory dependence is to invoke age-dependent birth-and-death rates. In our modelling, spontaneous emissions correspond to immigrations. It is natural to expect the immigration process to depend on memory effects; so far, we have assumed that the immigration process is a homogeneous Poisson process. In Chapter 6 we have chosen the parameter v of the Poisson immigration process to be equal to the birth-rate per individual, and this yielded the correct photon statistics for radiation below threshold level. Further, we have seen in sections 7.1 through 7.3 that the constancy of the expected rate of stimulated emission always leads to bunching. Now we propose to relax this assumption; the motivation to do so is twofold: first, spontaneous emission, being part of the emission process, cannot be treated differently from the rest of the population evolution process; further, the persistence of bunching is essentially due to spontaneous emission, and non-Markov modelling of the same may be expected to reduce bunching. There are many ways open to characterize the process of spontaneous emissions; however, we shall choose a simple one by assuming that the time-intervals (in the coarse-grained scale) between spontaneous emissions are independent and identically distributed non-negative random variables. In other words, the spontaneous emissions form a renewal process.

With the modification of spontaneous emission as mentioned above, let us review the results relating to the model discussed in sections 7.1 and 7.2. The conditioning used to define $g(z_1, z_2, t)$ has to be modified, leading to a new definition:

$$g(z_1, z_2, t)$$

7.4.1
$$= \lim_{\Delta \to 0} E\left[z_1^{X(t)} z_2^{Y(t)} \mid X(0) = 1 > X(-\Delta), v \neq 0 \right].$$

If we assume that the common p.d.f. of the time-interval between two successive emissions is given by $f_I(.)$, then it is easy to see that instead of 7.1.3 we obtain

$$g(z_1, z_2, t)$$

7.4.2
$$= g_1(z_1, z_2, t)\left[\bar{F}_I(t) + \int_0^t f_I(u) g(z_1, z_2, t - u)\, du \right],$$

an equation that can be used to generate the conditional moments of $X(t)$ and $Y(t)$.

If we define $M_k(t)$ by

7.4.3
$$M_k(t) = \left. \frac{\partial^k g(z_1, z_2, t)}{\partial z_1^k} \right|_{z_1 = z_2 = 1}$$

then we obtain from 7.4.2

7.4.4 $\qquad M_1(t) = {}_IM_1(t) + \int_0^t f_I(u)M_1(t-u)\,du$

$$M_2(t) = {}_IM_2(t) + 2_IM_1(t)\int_0^t f_I(u)M_1(t-u)\,du$$

7.4.5 $\qquad\qquad + \int_0^t f_I(u)M_2(t-u)\,du.$

Equation 7.4.4 is amenable to Laplace transformation. Thus taking the Laplace transform of both sides of 7.4.4, we obtain

7.4.6 $\qquad M_1^*(s) = {}_IM_1^*(s)/[1 - f_I^*(s)]$

from which it follows by the use of the Tauberian theorem that

7.4.7 $\qquad M_1 = \lim_{t\to\infty} M_1(t) = {}_IM_1^*(0)r_I$

where r_I is the rate of spontaneous emissions defined by

7.4.8 $\qquad 1/r_I = \int_0^\infty tf_I(t)\,dt.$

In a similar we we can obtain an explicit formula for M_2.

We next proceed to indicate the method of obtaining the product densities of the detection process. We take 7.2.1 as it is and modify the definition of $_0h_1(.)$ by a change of conditioning given by

7.4.9 $\qquad {}_0h_1(t) = \lim_{\Delta,\Delta'\to 0} Pr\{Y(t+\Delta) - Y(t) = 1 > X(-\Delta'), v \neq 0\}/\Delta.$

In a similar way the conditioning on the r.h.s. of 7.2.4 is modified. Thus all the statements and results pertaining to $_Ih_1(.)$ and $_Ih_2(\ ,\ .)$ are valid in our present model. Hence it is sufficient if we obtain the two product densities $_0h_1(.)$ and $_0h_2(.,\ )$.

To obtain the product density $_0h_1(.)$, we note that the photon that is detected by the detector may be the one that is produced in a population process generated by the one that is spontaneously emitted initially or by another. Using the fact that the epochs of spontaneous emissions form a renewal process, we obtain

7.4.10 $\qquad {}_0h_1(t) = {}_Ih_1(t) + \int_0^t h_I(u)_Ih_1(t-u)\,du,$

where $h_I(.)$ is the renewal density of the renewal process of spontaneous emissions. On taking Laplace transforms we obtain

7.4.11 $\qquad _0h_1^*(s) = {}_1h_1^*(s)(1 + h_I^*(s)).$

The limit of $_0h_1(t)$ as t tends to infinity is obtained by the use of the Tauberian theorem. The first term contributes zero, a result which can be seen even otherwise by appealing to the property of the branching process when $q < 1$. The second term contributes a non-zero factor since the renewal density $h_s(.)$ tends to the constant r_I. Thus we have

7.4.12 $\qquad \lim_{t \to \infty} {}_0h_1(t) = r_{I\,I}h_1^*(0).$

An expression for the second-order product density $_0h_2(.,.)$ is obtained by the use of combinatorial arguments. The two detections at t_1 and t_2 arise from the population generated by two different photons or a single photon. Analyzing these further, we obtain

$$_0h_2(t_1, t_2) = {}_1h_2(t_1, t_2) + \int_0^{\min(t_1,t_2)} h_I(u)_I h_2(t_1 - u, t_2 - u)\, du$$

$$+ \int_0^{t_1} h_I(u)_I h_1(t_1 - u)\, du \int_u^{t_2} h_I(v - u)_I h_1(t_2 - v)\, dv$$

7.4.13 $\qquad$
$$+ \int_0^{t_2} h_I(u)_I h_1(t_2 - u)\, du \int_u^{t_1} h_I(v - u)_I h_1(t_1 - v)\, dv$$

which on taking Laplace transform reduces to

$$_0h_2^*(s_1, s_2) = {}_1h_2^*(s_1, s_2)[1 + h_I^*(s_1 + s_2)]$$

7.4.14 $\qquad + {}_1h_1^*(s_1)_I h_1^*(s_2)h_I^*(s_1 + s_2)[h_I^*(s_1) + h_I^*(s_2)].$

Further reduction of the above formula is not possible for a general $h_I^*(.)$. It is easy to check that the above formula reproduces 7.2.15 by putting $h_I^*(s) = v/s$.

The special case when both $f(.)$ and $f_I(.)$ correspond to two-stage evolution is interesting, for such a model is versatile enough to contain features like statistics of thermal radiation on the one hand and anti-bunched and sub-Poissonian photon statistics on the other. Hence we provide a detailed description of the model in the following sections.

7.5 Two-stage model of evolution and emission: cavity population characteristics

In this model, the evolution of the field is taken to be a birth, death and immigration process. The field-detector interaction is modelled as an emigration process with a constant rate η per photon. The death (cavity absorption) rate is taken to be a constant equal to μ for the sake of simplicity in computations. The non-uniformity in birth rate is introduced as follows; the photon has a life-span at the end of which it is replaced by two, the life-span itself being the sum of two independently and exponentially distributed (positive) random variables with parameters α and λ. The immigration process is taken to be a renewal process where the interval between two successive immigrations is again the sum of two independently and exponentially distributed random variables with parameters β and v. It is to be noted that this model is different from the one introduced in section 7.3b.

There are some interesting limits; the limit as β and α tend to infinity yields the Markov model corresponding to constant rates of birth, death and immigration. If we take α finite and let β tend to infinity, then the process reduces to that discussed in section 7.1, provided we take $q = 1$ and replace η by $\mu + \eta$.

Next we observe that the two random variables whose sum constitutes the life-span can be interpreted to be the duration of the two phases. In the first phase whose duration is assumed to be an exponentially distributed random variable with parameter α, no birth is possible. On the termination of the first phase, the photon has a constant rate λ (probability per unit time) of being replaced by two photons, the second phase coming to an end on replacement. We can call the first phase the *dead (sterile) phase* and the second the *live (virile) phase*. The situation is depicted pictorially in Fig. 7.5.1 where photons are represented by wavy lines. Interpreted this way,

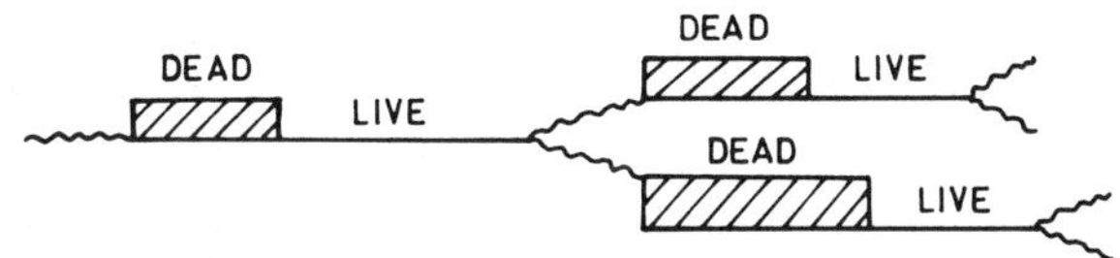

Fig. 7.5.1 Alternation of dead and live phases in the life-span of a photon

the model is close to the description of resonant fluorescence by Mandel (1979) who assumes that after each emission the atomic system makes a quantum jump, making further emissions not possible for some time. An

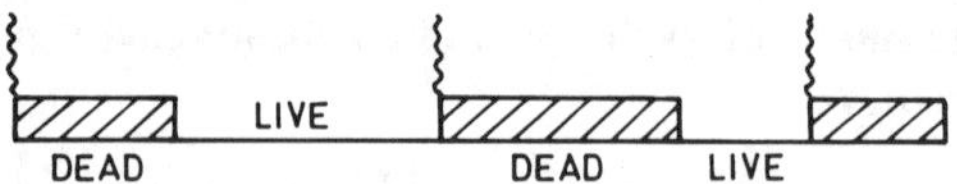

Fig. 7.5.2 Alternation of dead and live phases in the process of spontaneous emissions

exactly similar situation prevails for the immigration process. After each immigration (spontaneous emission) there is a dead phase whose duration is a random variable exponentially distributed with parameter β. The live phase which follows immediately after the termination of the dead phase comes to an end with the emission which now occurs at rate v. This can be represented pictorially as in Fig. 7.5.2. We call the resulting model the *two-phase model* or a *random dead-time model*.

The population evolution can also be viewed as a special case of the Bellman–Harris evolution introduced in section 7.1, provided we take out of our purview the supporting immigration process (or we take the limit as $\beta \to \infty$). To make contact with the Bellman–Harris type evolution, we note that the life-span corresponds to the time to death or the first birth. Observing that a birth is possible only after the termination of the dead phase and that death is possible in either phase, we obtain

$$7.5.1 \qquad f(x) = \mu\, e^{-(\alpha+\mu)x} + (\mu+\lambda)\alpha \int_0^x e^{-\alpha u - \lambda(x-u) - \mu x}\, du$$

or

$$7.5.2 \qquad f(x) = [\alpha(\mu+\lambda)\, e^{-(\mu+\lambda)x} - \lambda(\alpha+\mu)\, e^{-(\alpha+\mu)x}]/(\alpha-\lambda).$$

In the r.h.s. of 7.5.1, the first term corresponds to the termination of life-cycle due to death in sterile phase, while the second term corresponds to the termination of life-cycle due to death in virile phase or birth. With this interpretation, it is easy to identify the probability q:

$$7.5.3 \qquad q = \frac{\mu\lambda}{(\lambda+\mu)(\alpha+\mu)}.$$

Thus the two-phase model is a Bellman–Harris process with $f(\,.\,)$ defined by 7.5.2 and q by 7.5.3.

The model of birth-and-death process can also be viewed as a special case of age-dependent Kendall process if we interpret age-dependent probabilities in terms of phases as in section 4.3. In this case each cavity photon alternates between two phases, the alternation occurring at the time of birth. If we index the dead phase by 0 and live phase by 1, the movement of any photon in phases is a Markov chain with the following rates of transition

$$0 \overset{\alpha}{\to} 1, \qquad 1 \overset{\lambda}{\to} 0.$$

From Markov chain theory (see Example 2.2.2), we can arrive at the probability $p_{01}(x)$ of the photon remaining in phase 1 at age x, conditional on the phase being 0 at birth:

7.5.4 $\qquad p_{01}(x) = \alpha[1 - e^{-(\alpha + \lambda)x}]/(\alpha + \lambda).$

We can interpret the process in terms of age-dependent birth rate $\lambda(x)$ conditional upon survival up to age x:

7.5.5 $\qquad \lambda(x) = \alpha\lambda[1 - e^{-(\alpha + \lambda)x}]/(\alpha + \lambda).$

Thus the two-phase model or the random dead-time model can be interpreted as a special case of the Bellman–Harris process or an age-dependent Kendall process. In the Bellman–Harris model, the notion of life-span is rather hypothetical; on the other hand, in the age-dependent Kendall version, the life-span is actual life until absorption (death). The interesting fact that the cavity evolution can be described in terms of either model merely reinforces the non-Markov nature of the cavity evolution, thus according secondary importance to the notion of age or phase. However, the abstract and rather residuary nature of non-Markov evolution is interpreted by a concrete age/phase dependency which, besides rendering the model somewhat versatile, considerably simplifies the method of analysis.

7.5a Analysis of the model: generating functions

We call the photons in dead and live phases, respectively sterile and normal photons, and introduce the following notation:

$X_0(t)$: the number of sterile photons at time t;
$X_1(t)$: the number of normal photons at time t;
$W(t)$: the size of the cavity population at time t;
$\{Z(t)\}$: the two-state process representing the state of (cavity) immigration at time t; 0 and 1 denote respectively the dead and live phases.

We next introduce the generating functions of the conditional probability distribution of the cavity population by

7.5.6 $\qquad g_i(z_1, z_2, t) = E[z_1^{X_0(t)} z_2^{X_1(t)} \mid X_i(0) = 1 = W(0), v = 0] \qquad (i = 0, 1)$

7.5.7 $\qquad G_i(z_1, z_2, t) = E\left[z_1^{X_0(t)} z_2^{X_1(t)} \,\middle|\, W(0) = 0, Z(0) = i, v \neq 0\right] \qquad (i = 0, 1)$

and obtain the equations satisfied by the generating functions for arbitrary time $t > 0$ by analyzing the possible outcome in an infinitesimal interval $(0, \Delta)$. In the case of the functions $g_i(z_1, z_2, t)$ we have

$$g_0(z_1, z_2, t) = \left[1 - (\alpha + \mu + \eta)\Delta\right]g_0(z_1, z_2, t - \Delta)$$

7.5.8 $$+ \alpha\Delta g_1(z_1, z_2, t - \Delta) + (\mu + \eta)\Delta + o(\Delta)$$

$$g_1(z_1, z_2, t) = \left[1 - (\lambda + \mu + \eta)\Delta\right]g_1(z_1, z_2, t - \Delta)$$

7.5.9 $$+ \lambda\Delta\left[g_0(z_1, z_2, t - \Delta]^2 + (\mu + \eta)\Delta + o(\Delta).\right.$$

To obtain 7.5.8 we have used the fact that the sterile photon present at $t = 0$ can, over the interval $(0, \Delta)$,

    (i)   move into normal state with probability $\alpha\Delta + o(\Delta)$;

    (ii) be absorbed by the cavity with probability $\mu\Delta + o(\Delta)$;

    (iii) be removed by the detector with probability $\eta\Delta + o(\Delta)$;

    (iv) continue to be in the same state with the residual probability $1 - (\alpha + \mu + \eta)\Delta + o(\Delta)$.

Similar arguments lead to 7.5.9, the only difference being that a normal photon can produce a sterile photon over the interval $(0, \Delta)$, with the normal photon turning sterile with probability $\lambda\Delta + o(\Delta)$. Proceeding to the limit as $\Delta \to 0$, we find

7.5.10 $\qquad \dfrac{\partial g_0(z_1, z_2, t)}{\partial t} = -(\alpha + \mu)g_0(z_1, z_2, t) + \alpha g_1(z_1, z_2, t) + \mu$

7.5.11 $\qquad \dfrac{\partial g_1(z_1, z_2, t)}{\partial t} = -(\lambda + \mu)g_1(z_1, z_2, t) + \lambda\left[g_0(z_1, z_2, t)\right]^2 + \mu$

where we have used μ *in place of* $\mu + \eta$ for notational convenience. The initial conditions are given by

7.5.12 $\qquad g_i(z_1, z_2, 0) = z_{i+1} \qquad (i = 0, 1).$

To obtain the equations satisfied by the functions $G_i(z_1, z_2, t)$ $(i = 0, 1)$ we study the vector process $\{Z(t), X_0(t), X_1(t)\}$. We fix our attention on the time-interval $(0, \Delta)$ as before; for $i = 0$, the state of the cavity (state of immigration) which is in dead phase

    (i) moves to the live phase with probability $\beta\Delta + o(\Delta)$ due to the exponential nature of the life-time distribution;

    (ii) stays in dead phase with probability $1 - \beta\Delta + o(\Delta)$.

On the other hand, if the cavity is in live phase, then

   (i) there is a spontaneous emission (immigration) with probability $v\Delta + o(\Delta)$, in which case the emitted photon generates a cavity population over the time-interval (Δ, t) with corresponding generating function $g_0(z, t - \Delta)$, this being independent of the process $\{Z(t)\}$; or

   (ii) the cavity continues to be in live phase with residual probability $1 - v\Delta + o(\Delta)$.

Taking all these possibilities into account, we obtain

$$7.5.13 \qquad \frac{\partial G_0(z_1, z_2, t)}{\partial t} = -\beta G_0(z_1, z_2, t) + \beta G_1(z_1, z_2, t)$$

$$7.5.14 \qquad \frac{\partial G_1(z_1, z_2, t)}{\partial t} = -v G_1(z_1, z_2, t) + v G_0(z_1, z_2, t) g_0(z_1, z_2, t)$$

with the initial conditions

$$7.5.15 \qquad G_i(z_1, z_2, 0) = 1.$$

The equilibrium distribution of cavity photons is obtained by taking the limit as $t \to \infty$ of $G_0(z_1, z_2, t)$ or $G_1(z_1, z_2, t)$. It is indeed difficult to obtain explicit solutions for $g_i(z_1, z_2, t)$ or $G_i(z_1, z_2, t)$. However, the first moments can be obtained by appropriate differentiation of equations 7.5.10, 7.5.11, 7.5.13 and 7.5.14. Introducing the moments by

$$7.5.16 \qquad a_i^j(t) = \left. \frac{\partial g_i(z_1, z_2, t)}{\partial z_j} \right|_{z_1 = z_2 = 1}$$

$$7.5.17 \qquad b_i^{jk}(t) = \left. \frac{\partial^2 g_i(z_1, z_2, t)}{\partial z_j \partial z_k} \right|_{z_1 = z_2 = 1}$$

$$7.5.18 \qquad A_i^j(t) = \left. \frac{\partial G_i(z_1, z_2, t)}{\partial z_j} \right|_{z_1 = z_2 = 1}$$

$$7.5.19 \qquad B_i^{jk}(t) = \left. \frac{\partial^2 G_i(z_1, z_2, t)}{\partial z_j \partial z_k} \right|_{z_1 = z_2 = 1}$$

where $i = 0, 1$; $j, k = 1, 2$, the moments of the cavity population $W(t)$ can easily be expressed in terms of the above moments. Defining $a_i(t)$, $b_i(t)$, $A_i(t)$, $B_i(t)$ by

$$7.5.20 \qquad a_i(t) = E[W(t) \,|\, X_i(0) = 1 = W(0), v = 0]$$

$$7.5.21 \qquad b_i(t) = E[W(t)(W(t) - 1) \,|\, X_i(0) = 1 = W(0), v = 0]$$

7.5.22 $\quad A_i(t) = E[W(t)\,|\,W(0) = 0, z(0) = i, v \neq 0]$

7.5.23 $\quad B_i(t) = E[W(t)(W(t) - 1)\,|\,W(0) = 0, Z(0) = i, v \neq 0]$,

we note the following connecting relations:

7.5.24 $\quad a_i(t) = \sum_j a_i^j(t), \quad A_i(t) = \sum_j A_i^j(t)$

7.5.25 $\quad b_i(t) = \sum_{jk} b_i^{jk}(t), \quad B_i(t) = \sum_{jk} B_i^{jk}(t)$.

We next note that the first two moments of the equilibrium distribution are simply given by $A_i(\infty)$ and $B_i(\infty)$.

7.5b Moment structure

To obtain the moments, we differentiate equations 7.5.10 and 7.5.11 appropriately and obtain

7.5.26 $\quad \dfrac{da_0^j(t)}{dt} = -(\alpha + \mu)a_0^j(t) + \alpha a_1^j(t)$

7.5.27 $\quad \dfrac{da_1^j(t)}{dt} = -(\lambda + \mu)a_1^j(t) + 2\lambda a_0^j(t)$

7.5.28 $\quad \dfrac{db_0^{jk}(t)}{dt} = -(\alpha + \mu)b_0^{jk}(t) + \alpha b_1^{jk}(t)$

7.5.29 $\quad \dfrac{db_1^{jk}(t)}{dt} = -(\lambda + \mu)b_1^{jk}(t) + 2\lambda b_1^{jk}(t) + 2\lambda a_0^j(t)a_0^k(t)$,

where the initial conditions are given by

7.5.30 $\quad a_0^1(0) = 1, \quad a_1^2(0) = 1, \quad a_1^1(0) = a_0^2(0) = 0$

7.5.31 $\quad b_i^{jk}(0) = 0, \quad i = 0, 1$.

The functions $a_i^j(t)$ ($j = 1, 2, i = 0, 1$) can be obtained explicitly. If $a_i^{j*}(s)$ is the Laplace transform of $a_i^j(t)$, then using 7.5.30, we obtain

$$a_0^{1*}(s) = (\lambda + \mu + s)/D(s), \qquad a_0^{2*}(s) = \alpha/D(s).$$

7.5.32 $\quad a_1^{1*}(s) = 2\lambda/D(s), \qquad\qquad a_1^{2*}(s) = (\alpha + \mu + s)/D(s)$,

where

7.5.33 $\quad D(s) = (\alpha + \mu + s)(\lambda + \mu + s) - 2\lambda\alpha$.

Inverting the transforms, we obtain

$$a_0^1(t) = \{(\lambda + \mu + s_1)e^{s_1 t} - (\lambda + \mu + s_2)e^{s_2 t}\}/(s_1 - s_2)$$

$$a_1^1(t) = \frac{2\lambda}{\alpha}\, a_0^2(t) = 2\lambda(e^{s_1 t} - e^{s_2 t})/(s_1 - s_2)$$

7.5.34 $\quad a_1^2(t) = \{(\alpha + \mu + s_1)e^{s_1 t} - (\alpha + \mu + s_2)e^{s_2 t}\}/(s_1 - s_2)$

where s_1 and s_2 are given by

7.5.35 $\quad 2s_{1,2} = -(2\mu + \lambda + \alpha) \pm \sqrt{(\lambda + \alpha)^2 + 4\lambda\alpha}\,.$

In a similar way the functions $b_i^{jk}(t)$ can be obtained, but it is sufficient for our purposes if we obtain the Laplace transform solution. After some straightforward computations, we have for $j, k = 1, 2$

7.5.36 $\quad b_0^{jk}(s) = 2\alpha\lambda L^{jk}(s)/D(s)$

7.5.37 $\quad b_1^{jk*}(s) = 2\lambda(\alpha + \mu + s)L^{jk}(s)/D(s),$

where

7.5.38 $\quad L^{jk}(s) = \displaystyle\int_0^\infty e^{-st} a_0^j(t)a_0^k(t)\, dt.$

It is rather tedious to evaluate the integral on the r.h.s. of 7.5.38. We can convert it into an appropriate contour integral by the use of the inversion formula for the function $a_0^j(t)$. After some computations, we obtain

$$L^{11}(s) = \left\{ \frac{(\lambda + \mu + \xi_1)(\lambda + \mu + s - \xi_1)}{D(\xi_1)} \right.$$

7.5.39 $$\left. - \frac{(\lambda + \mu + \xi_2)(\lambda + \mu + s - \xi_2)}{D(\xi_2)} \right\} \bigg/ (\xi_2 - \xi_1)$$

7.5.40 $\quad L^{12}(s) = L^{21}(s) = \alpha\left\{ \dfrac{\lambda + \mu + \xi}{D(\xi_1)} - \dfrac{\lambda + \mu + \xi_2}{D(\xi_2)} \right\} \bigg/ (\xi_2 - \xi_1)$

7.5.41 $\quad L^{22}(s) = 4\lambda^2\left[\dfrac{1}{D(\xi_1)} - \dfrac{1}{D(\xi_2)} \right] \bigg/ (\xi_2 - \xi_1),$

where

7.5.42 $\quad 2\xi_{1,2} = (\lambda + \alpha + 2\mu + 2s) \mp \sqrt{(\alpha + \lambda)^2 + \lambda\alpha}\,.$

We next differentiate 7.5.13 and 7.5.14 w.r.t. z_1 and z_2 appropriately, to obtain for $j = 1, 2$

$$7.5.43 \qquad \frac{\mathrm{d}A_0^j(t)}{\mathrm{d}t} = -\beta A_0(t) + \beta A_1^j(t)$$

$$7.5.44 \qquad \frac{\mathrm{d}A_1^j(t)}{\mathrm{d}t} = -vA_1^j(t) + vA_0^j(t) + va_0^j(t),$$

with the initial conditions

$$7.5.45 \qquad A_0^j(0) = A_1^j(0) = 0.$$

The Laplace transform solution of 7.5.43 and 7.5.44 can be readily formed:

$$7.5.46 \qquad A_0^{j*}(s) = \frac{\beta v a_0^{j*}(s)}{s(s + \beta + v)}$$

$$7.5.47 \qquad A_1^{j*}(s) = \frac{v(s + \beta)a_0^{j*}(s)}{s(s + \beta + v)}.$$

The expected value of the equilibrium cavity population can be obtained by use of the Tauberian theorem:

$$\lim_{t \to \infty} A_i^j(t) = \lim_{s \to 0} s A_i^{j*}(s)$$

$$7.5.48 \qquad\qquad = \frac{\beta v}{\beta + v} a_0^{j*}(0).$$

The constants $a_0^{j*}(0)$ can be readily obtained:

$$7.5.49 \qquad a_0^{1*}(0) = (\lambda + \mu)/D(0)$$

$$7.5.50 \qquad a_0^{2*}(0) = \alpha/D(0).$$

Thus the mean size of the cavity population in equilibrium is given by

$$\lim_{t \to \infty} A_i(t) = \lim_{t \to \infty} \sum_j A_i^j(t)$$

$$7.5.51 \qquad\qquad = (\lambda + \mu + \alpha)\beta v/D(0)(\beta + v).$$

To obtain the second moments, we differentiate 7.5.13 and 7.5.14 twice w.r.t. z_1 and z_2 appropriately, to obtain:

$$7.5.52 \qquad \frac{\mathrm{d}B_0^{jk}(t)}{\mathrm{d}t} = -\beta B_0^{jk}(t) + \beta B_1^{jk}(t)$$

$$\frac{dB_1^{jk}(t)}{dt} = -vB_1^{jk}(t) + vB_0^{jk}(t)$$

$$7.5.53 \qquad + vb_0^{jk}(t) + v\left[A_0^k(t)a_0^j(t) + A_0^j(t)a_0^k(t)\right]$$

with the initial conditions

$$7.5.54 \qquad B_1^{jk}(0) = B_0^{jk}(0) = 0.$$

As before, we obtain the Laplace transform solution:

$$7.5.55 \qquad B_0^{jk*}(s) = \frac{\beta v}{s(s+\beta+v)}\left[b_0^{jk*}(s) + M^{jk}(s)\right]$$

$$7.5.56 \qquad B_1^{jk*}(s) = \frac{(\beta+s)}{s(s+\beta+v)}\left[b_0^{jk*}(s) + M^{jk}(s)\right]$$

where

$$7.5.57 \qquad M^{jk}(s) = \int_0^\infty e^{-st}\left[A_0^j(t)a_0^k(t) + A_0^k(t)a_0^j(t)\right]dt.$$

The second moments of the equilibrium cavity population are obtained again by use of the Tauberian theorem:

$$\lim_{t\to\infty} B_i^{jk}(t) = \lim_{s\to 0} sB_i^{jk*}(s)$$

$$7.5.58 \qquad = \frac{\beta v}{\beta+v}\left[b_0^{jk*}(0) + M^{jk}(0)\right].$$

The functions $b_0^{jk*}(s)$ $(j,k=1,2)$ are given by 7.5.36 through 7.5.42. Thus it is sufficient to calculate the functions $M^{ij}(0)$. It is rather tedious to evaluate 7.5.57 directly. We note that

$$7.5.59 \qquad M^{jk}(0) = \frac{1}{2\pi i}\int_{-i\infty}^{+i\infty}\left[A_0^{j*}(\xi)a_0^{k*}(-\xi) + A_0^k(\xi)a_0^{j*}(-\xi)\right]d\xi.$$

Evaluating the integrals on the r.h.s. of 7.5.59 by converting them into appropriate closed contour integrals, we obtain

$$7.5.60 \qquad M_0^{11}(0) = -K_1(\alpha,\beta) + (\lambda+\mu)K_2(\alpha,\beta)$$

$$7.5.61 \qquad M_0^{21}(0) = M_0^{12}(0) = \alpha K_2(\alpha,\beta)$$

$$7.5.62 \qquad M_0^{22}(0) = \alpha^2 K_2(\alpha,\beta)/(\lambda+\mu)$$

$$7.5.63 \qquad 1/K_1(\alpha,\beta) = (\alpha+\lambda+2\mu)\left[(\beta+v)^2 + (\beta+v)(\alpha+\lambda+2\mu) + s_1s_2\right]/\beta v$$

$$7.5.64 \qquad K_2(\alpha, \beta) = (\lambda + \mu)\big[(\alpha + \lambda + 2\mu)^2 - s_1 s_2$$
$$+ (\beta + \nu)(\alpha + \lambda + 2\mu)\big]/s_1^2 s_2^2.$$

7.5c Behaviour of the second moment: special cases

We now examine the behaviour of the second moment of the equilibrium cavity population under various approximations. We first take the limit as $\beta \to \infty$. As observed earlier, this corresponds to a special case of the age-dependent model of the branching type considered in section 7.1 for the choice $f(\,.\,)$, and q given by 7.5.2 and 7.5.3. The expected value of the cavity population in equilibrium follows from 7.5.51:

$$7.5.65 \qquad A(\infty) = A_i(\infty) = \nu(\lambda + \mu + \alpha)/D(0).$$

The second factorial moment of the equilibrium cavity population follows from 7.5.58 and is given by

$$7.5.66 \qquad B(\infty) = [b_0^*(0) + M(0)]\nu,$$

where $b_0^*(0)$ and $M(0)$ are given by

$$7.5.67 \qquad b_0^*(0) = \sum_{j,k} b_0^{jk}{}^*(0), \quad M(0) = \sum_{jk} M_0^{jk}(0)$$

$$b_0^{11}{}^*(0) = \alpha\lambda[(\lambda + \mu)^2 + D(0)]/E, \quad b_0^{21}{}^*(0) = \lambda\alpha^2(\lambda + \mu)/E$$

$$b_0^{22}{}^*(0) = \lambda\alpha^3/E, \quad E = [D(0)]^2(\lambda + \alpha + 2\mu)$$

$$7.5.68 \qquad M_0^{11}(0) = \nu(\lambda + \mu)^2/[D(0)]^2,$$

$$M_0^{12}(0) = \alpha\nu(\lambda + \mu)/[D(0)]^2$$

$$M_0^{22}(0) = \alpha^2\nu/[D(0)]^2.$$

Using the above expressions for $b_0^{ij}{}^*(0)$ and $M_0^{ij}(0)$, we finally obtain the following formula for the second factorial moment:

$$7.5.69 \qquad B(\infty) = (A(\infty))^2 + \nu\alpha\lambda\{(\lambda + \mu + \alpha)^2 + D(0)\}/E.$$

The cavity population statistics will correspond to those of thermal photons, at least up to second order, provided that

$$7.5.70 \qquad B(\infty) = 2(A(\infty))^2.$$

The above condition reduces to the following choice of ν:

$$7.5.71 \qquad \nu = \frac{\lambda\alpha}{\lambda + \alpha + \mu}\,[1 - \lambda\alpha/(\lambda + \alpha + \mu)(\lambda + \alpha + 2\mu)].$$

Thus the non-Markov evolution of a cavity population can still lead to thermal behaviour to second order, provided that v is chosen according to 7.5.71. We shall examine below (section 7.6) the possibility of reproducing the Lorentzian spectrum.

If we further proceed to the limit as $\alpha \to \infty$, the model reduces to the Markov model dealt with in Chapter 6 and the choice 7.5.71 leads to $v = \lambda$.

We now return to the general case when β is comparable to α. As the expressions for the moments are rather unwieldy, we set $\alpha = \beta$. From a physical point of view, there is no particular reason why the dead time following the spontaneous and stimulated emissions would be differently distributed. The first two moments of the equilibrium cavity population are now given by

$$7.5.72 \qquad A(\infty) = \alpha v(\lambda + \mu + \alpha)/[(\alpha + v)D(0)]$$

$$7.5.53 \qquad B(\infty) = \alpha v[b_0^*(0) + M(0)]/(\alpha + v),$$

where $b_0^*(0)$ specified by 7.5.67 and 7.5.68 is still valid for the present case. Only the explicit expression for $M(0)$ need to be specified. At this stage, it is worthwhile to set $v = \lambda$; using the general formula given by 7.5.60 through 7.5.62, we finally obtain

$$7.5.74 \qquad B(\infty) = \frac{2(\alpha\lambda)^2[2(\alpha + \lambda + \mu)^4 + D(0)(\alpha + \lambda)(\alpha + \lambda + \mu)]}{(\alpha + \lambda)(\alpha + \lambda + 2\mu)[(2(\alpha + \lambda)(\alpha + \lambda + \mu) + D(0)][D(0)]^2}.$$

We next examine the possibility of the photon statistics being anti-bunched. This is possible if the quantity $\mathscr{B}$ defined by

$$7.5.75 \qquad \mathscr{B} = B(\infty)/[A(\infty)]^2$$

is less than 1. Using 7.5.74 and 7.5.72, we note that $\mathscr{B}$ is given by

$$7.5.76 \qquad \mathscr{B} = \frac{2(\alpha + \lambda)2(\alpha + \lambda + \mu)^2 + D(0)(\alpha + \lambda)/(\alpha + \lambda + \mu)}{(\alpha + \lambda + 2\mu)[2(\alpha + \lambda + \mu)(\alpha + \lambda) + D(0)]}.$$

There are several choices of α, μ for which $\mathscr{B}$ is less than 1. For instance, for $\alpha = \lambda, \mu = 3\lambda$ we find $\mathscr{B} = 0.82$. Likewise for $2\alpha = \lambda, \mu = 2\lambda, \mathscr{B} = 0.88$. When $\mathscr{B}$ is less than 1, the statistics also exhibit sub-Poissonian behaviour. Such anti-bunched or sub-Poissonian behaviour of photon statistics is generally attributed to non-classical nature of the resulting radiation. It should be noted that the model of cavity population of photons that we have proposed follows essentially from the usual (fully) quantum mechanical density matrix treatment; the only difference lies in the manner of specifying the evolution equation in terms of coarse-grained time. The non-

Markov nature of the evolution arises essentially from the averaging process over coarse-grained time scale.

As we have observed earlier, it is indeed possible for the resulting cavity photon statistics to possess bunched/super Poissonian character for appropriate choice of the parameters α, λ, μ. For instance, the values (i) $\alpha = \lambda$, $\mu = 2\lambda$ and (ii) $4\alpha = \lambda$, $\mu = \lambda$ lead to bunched/super-Poissonian behaviour of the resulting radiation.

7.6   Two-stage model: detection process

We are now in a position to study the photodetection process. Since the detection process is essentially a destructive scheme of sampling the cavity photon population, we have modelled the process as an emigration process with a constant rate η per unit time. The constancy of the rate is again a characteristic of the ideal nature of the detection process. We are interested in the random process $N(t_0, t) = Y(t_0 + t) - Y(t_0)$ representing the number of photons detected over the time-interval $(t_0, t_0 + t)$. In our model the process becomes stationary, and hence the distributional characteristics of the process $N(t_0, t)$ are independent of t_0. It is convenient as in the previous section to introduce the generating function of $N(t_0, t)$ or $N(t)$ by (for $i = 0, 1$)

7.6.1 $\qquad g_i(z, t) = E\left[z^{N(t)} \,\middle|\, X(0) = 1 - i, \, T(0) = i, \, v = 0\right]$

7.6.2 $\qquad G_i(z, t) = E\left[z^{N(t)} \,\middle|\, X(0) = 0 = Y(0), \, Z(0) = i, \, v \neq 0\right]$.

If we use the same arguments as in section 3 and note that photons of either type can be removed by destructive sampling, we obtain the following equations:

7.6.3 $\qquad \dfrac{\partial g_0(z, t)}{\partial t} = -(\alpha + \mu + \eta)g_0(z, t) + ag_1(z, t) + \mu + \eta z$

7.6.4 $\qquad \dfrac{\partial g_1(z, t)}{\partial t} = -(\lambda + \mu + \eta)g_1(z, t) + \lambda[g_0(z, t)]^2 + \mu + \eta z$

7.6.5 $\qquad \dfrac{\partial G_0(z, t)}{\partial t} = -\beta G_0(z, t) + \beta G_1(z, t)$

7.6.6 $\qquad \dfrac{\partial G_1(z, t)}{\partial t} = -vG_1(z, t) + vG_0(z, t)g_0(z, t)$.

The initial conditions are given by

7.6.7 $\qquad g_i(z, 0) = z, \quad G_i(z, 0) = 1 \quad (i = 0, 1)$.

The moments of $N(t)$ can be obtained by differentiating the above equations w.r.t. z at $z = 1$ and solving the resulting set of equations. The structure of the differential equations is the same as in section 7.5. We shall not proceed further in this direction.

There is an alternative line of approach leading to the correlational structure of the point process of detection. The detection process can be characterized in terms of the sequence of product densities. For the model under discussion, these are conditional product density/intensity functions and are defined by

7.6.8 $\qquad f_1(t) = \lim_{\Delta \to 0} Pr\{N(t + \Delta) - N(t) = 1 \,|\, \text{cavity in eqm. initially}\}/\Delta$

$$f_2(t_1, t_2) = \lim_{\Delta, \Delta' \to 0} Pr\{N(t_1 + \Delta) - N(t_1) = 1, N(t_2 + \Delta') - N(t_2)$$

7.6.9 $\qquad\qquad\qquad\qquad\qquad = 1 \,|\, \text{cavity in eqm. initially}\}/\Delta\Delta',$

with higher-order functions defined in a similar manner. From the construction of our mode, it is clear that the process is stationary, and consequently we have

7.6.10 $\qquad f_1(t) = \text{a constant}$

7.6.11 $\qquad f_2(t_1, t_2) = \text{a function of } |t_2 - t_1|$

$$= h_{sty}(|t_2 - t_1|).$$

If we confine our attention to second-order characteristics, we need only to identify the constant on the r.h.s. of 7.6.10 and obtain an explicit expression for the function $h_{sty}(\,.\,)$.

We first obtain the product densities corresponding to different conditions introduced as in earlier sections and then express the functions f_1 and f_2 in terms of these conditional product densities. Accordingly we define

$$_ih_1^i(t) = \lim_{\Delta \to 0} Pr\{N(t + \Delta) - N(t) = 1 \,|$$

7.6.12 $\qquad\qquad\qquad X(0) = 1 - i, Y(0) = i, v = 0\}/\Delta \quad i = 0, 1$

$$h_1(t) = \lim_{\Delta \to 0} Pr\{N(t + \Delta) - N(t) = 1 \,|$$

7.6.13 $\qquad\qquad\qquad X(0) = Y(0) = 0, v \neq 0\}/\Delta$

$$h_1^E(t) = \lim_{\Delta \to 0} Pr\{N(t + \Delta) - N(t) = 1 \mid X(0) = 0 = Y(0),$$

7.6.14
$$\text{cavity in eqm., } v \neq 0\}/\Delta.$$

We first note that $h_1(.)$ and $f_1(.)$ are directly related:

7.6.15 $$f_1(t) = \lim_{\tau \to \infty} h_1(\tau).$$

Using the definition 7.6.13, we obtain

7.6.16 $$f_1(t) = \lim_{\tau \to \infty} A(\tau)\eta$$

$$= \beta v \eta (\lambda + \mu + \alpha)/[(\beta + v)D(0)].$$

Interpreting the r.h.s. of 7.6.12 and 7.6.14 as the conditional rates of detection, we obtain

7.6.17 $$_1 h_1^i(t) = a_i(t)$$

7.6.18 $$h_1^E(t) = (p_0 A_0(t) + p_1 A_1(t)),$$

where p_0 and p_1 are the equilibrium probabilities of the process $\{Z(t), t \geq 0\}$. Observing that the process $\{Z(t), t \geq 0\}$ is a two-state Markov chain with the rates of transition

$$0 \xrightarrow{\beta} 1, \qquad 1 \xrightarrow{v} 0,$$

we find

7.6.19 $$p_0 = v/(\beta + v) \qquad p_1 = \beta/(\beta + v).$$

Using the Laplace transform solution 7.5.46, 7.5.47 and the relations 7.5.24 and 7.5.25, we obtain after some calculations

7.6.20
$$h_1^E(t) = \frac{\beta \eta v}{(\beta + v)(s_1 - s_2)} \left\{ (\alpha + \mu + \lambda + s_1) \frac{e^{s_1 t} - 1}{s_1} \right.$$
$$\left. - (\alpha + \mu + \lambda + s_2) \frac{e^{s_2 t} - 1}{s_2} \right\}$$

7.6.21
$$_1 h_1^0(t) = \eta\{(\alpha + \mu + \lambda + s_1) e^{s_1 t}$$
$$- (\alpha + \mu + \lambda + s_2) e^{s_2 t}\}/(s_1 - s_2)$$

7.6.22
$$_1 h_1^i(t) = \eta\{(\alpha + \mu + 2\lambda + s_1) e^{s_1 t}$$
$$- (\alpha + \mu + 2\lambda + s_2) e^{s_2 t}\}/(s_1 - s_2).$$

Finally, to obtain $h_{sty}(t)$ $(t > 0)$ we make full use of the branching nature of the cavity population evolution. The arguments are typically combinatorial in nature and run as follows. The function $h_{sty}(t)$ represents the expected value of the product of the rate of one detection at the origin when the cavity is in equilibrium and the rate of one detection at the epoch t. The photon that is detected can be from either (i) the population generated by one of the photons present at the time-origin or (ii) the population generated by a photon which is itself a result of spontaneous emission at some epoch between 0 and t. In the case of eventuality (i) the photon could be of type 0 (sterile) or 1 (virile). In the case of eventuality (ii), the probabilities corresponding to the rates can be multiplied. Thus we have

$$7.6.23 \qquad h_{sty}(t) = \eta[B^{12} + B^{11})_0 h_1^0(t) + (B^{22} + B^{21})_0 h_1^1(t)] + h_1(\infty) h_1^E(t),$$

where

$$7.6.24 \qquad B^{jk} = \lim_{t \to \infty} B_i^{jk}(t)$$

and hence given by the r.h.s. of 7.5.58.

As observed earlier, a measure of bunching $\mathscr{B}$ can be defined through the statistics of the detection process:

$$7.6.25 \qquad \mathscr{B} = h_{sty}(0)/h_{sty}(\infty).$$

It is easy to see that this definition is consistent with the earlier definition in terms of the second factorial moment given by 7.5.71.

Next we study the limit as $\beta \to \infty$. As we have seen earlier in section 7.4, $\mathscr{B} = 2$, provided we choose v through equation 7.5.71. An interesting question arises whether the spectrum of the detection process can be Lorentzian. To answer this question, we observe that in this case 7.6.23 reduces to

$$h_{sty}(t) = (h_1(\infty))^2 + \frac{\eta}{s_1 - s_2}$$

$$\times \left\{ e^{s_1 t}\left[(\alpha + \mu + \lambda + s_1)\left\{ \frac{vh_1(\infty)}{s_1} + B(\infty)\eta \right\} + \eta\lambda(B^{21} + B^{22}) \right] \right.$$

$$7.6.26 \qquad \left. - e^{s_2 t}\left[(\alpha + \mu + \lambda + s_2)\left\{ \frac{vh_1(\infty)}{s_2} + B(\infty)\eta \right\} + \eta(B^{21} + B^{22}) \right] \right\}.$$

We next note if the coefficient of $e^{s_2 t}$ is zero for some choice of μ and α, then

7.6.26 reduces to

7.6.27 $\qquad h_{sty}(t) = (h_1(\infty))^2[1 + e^{s_1 t}].$

Thus the resulting radiation is thermal with a Lorentzian profile. The condition for the vanishing of the coefficient of $e^{s_2 t}$ can be written as

$$7.6.28 \qquad v(\mu + \lambda + \alpha)(\mu + \lambda + \alpha + s_2) + \frac{\lambda^2 \alpha_\lambda^2}{\lambda + \alpha + 2\mu} = 0.$$

From 7.5.35, we observe that $s_2 < s_1 < 0$ and $\mu + \lambda + \alpha + s_2 < 0$. If at this stage we use 7.5.71, we can reduce 7.6.28 to

$$7.6.29 \qquad 8(\lambda + \alpha + \mu)[\mu(\lambda + \alpha + 2\mu) - \lambda\alpha] - 12\lambda\alpha\mu + \frac{4\lambda^2\alpha^2}{\lambda + \alpha + \mu} = 0,$$

from which we conclude that there are many choices of positive values of λ, α, and μ satisfying 7.6.29. Thus the non-Markov evolution of the cavity population with the parameters subject to the constraints 7.5.71 and 7.6.29 produces a thermal stream with a Lorentzian spectrum. It should be noted that the result is true only to second order since $h_{sty}(.)$ brings out only second-order characteristics of the detection process.

We finally return to the general case when β is finite. If we set $\beta = \alpha$ as in section 7.5 then $\mathscr{B}$ is given by 7.5.76, and it is possible to have several choices of α, λ and μ that lead to the sub-Poissonian/anti-bunched character of the resulting photon statistics.

Additional notes

Earlier work on non-Markov modelling is based on an extension of the Pauli–Master equation in which the transition rates are assumed to be time-dependent. Bonifacio *et al.* (1971) studied the problem of super-radiance using such a non-Markov approach. Huang *et al.* (1981) dealt with popular equations for quantum systems with dissipative mechanisms; these authors introduce memory effects by generalizing the transport coefficients. The type of memory effects dealt with in this chapter is different. The phenomenon of anti-bunching was first observed in resonance fluorescence by Kimble, Dagenais and Mandel (1977, 1978); other mechanisms of generating light with such properties were suggested by Mandel (1979), Loudon (1980), Short and Mandel (1983), Teich, Saleh and Stoller (1983), Walker and Jakeman (1985), and Teich and Saleh (1985).

The surveys by Paul (1982) and Teich and Saleh (1987) give a detailed account of photon anti-bunching from both the theoretical and the experimental points of view. The analysis and results presented in this chapter are adapted from Srinivasan and Vasudevan (1986a) and Srinivasan (1986c, 1987).

8. Non-Markov cavity radiation – II: Kendall process

In Chapter 7 we dealt with various models that come broadly under the Bellman–Harris type of age-dependent evolution supported by an immigration process. In this chapter we consider models that correspond to age-dependent evolution (of the Kendall type) in which the life span of a photon is the actual life-span. More specifically we deal with models with birth rates of certain special age-dependency that corresponds to the phase-type evolution introduced in Chapter 4. We first deal with a three-phase evolution and show that the resulting radiation can under certain conditions reduce to a thermal stream with a Lorentzian profile (at least up to second order). We then generalize the model and deal with an asymmetric phase evolution; by an appropriate choice of the parameters of the distribution of the life-span of the phases, we show that the resulting radiation corresponds to a thermal stream with a profile that corresponds to the superposition of Lorentzian profiles of different widths. Finally we show that an appropriate choice of the immigration process leads to anti-bunching and sub-Poissonian photon statistics.

8.1 Modified Kendall evolution: phase approach

We assume that the cavity photons evolve according to a modified Kendall process of birth, death and immigration. The death (cavity absorption) and immigration (spontaneous emission) rates are assumed to be constants, respectively equal to μ and v. The birth rate is assumed to be specified by the age-dependent function $\lambda(x)$ given by

8.1.1 $\qquad \lambda(x) = \alpha\, e^{-\lambda x}(\lambda x)^{n-1}/(n-1)!$

where α and λ are two constants having the dimensions of reciprocal of time and n is any positive integer. We will be mostly concerned with the cases $n = 2$ and 3; later on, we shall outline the method of analysis for any arbitrary value of n. In passing we note that $n = 1$ is rather uninteresting from our point of view of modelling cavity population. The photodetection process corresponds to emigration at a constant rate η. As discussed in

Chapter 4, the form 8.1.1 is highly suggestive of a possible phase transformation by the cavity photons. For $n = 2$ the life span of any photon conditional upon its survival (from death and detection) can be thought of as the sum of three phases. The first two phases have durations T_1 and T_2 that are independent and exponentially distributed with the same parameter λ. In the first phase, the photons passively interact with the cavity, producing no additional photons at all. In the second phase the photons are active and each photon produces an additional photon at a constant rate α. The third phase is assumed to have indefinite span during which no births are possible. Although the span of the third phase is indefinite, the photon will eventually be absorbed or detected. If we revert to the description in terms of age, a photon of age x is in

(i) phase 1 with probability $e^{-\lambda x}$;

(ii) phase 2 with probability $e^{-\lambda x}\lambda x$; and

(iii) phase 3 with the residual probability $1 - e^{-\lambda x}(1 + \lambda x)$.

Thus the probability that a photon of age x at time t conditional upon its survival produces a photon of age 0 in the time-interval $(t, t + \Delta)$ is

$$e^{-\lambda x}\lambda x\alpha\Delta + o(\Delta),$$

so that the rate is given by 8.1.1 when $n = 2$. As observed earlier in Chapter 4, the process is different from the age-dependent process, although the marginal birth rate coincides with that of the corresponding age-dependent process. The study of cavity evolution through phases is straightforward and follows from the general analysis of population growth presented in Chapter 4. However, here we are primarily interested in the characterization of the photocount distribution with special reference to the correlation of photocounts. Let $X_1(t)$, $X_2(t)$ and $X_3(t)$ represent the size of the cavity population at time t in phases 1, 2 and 3 respectively; let $W(t)$ represent the total size of the population. We introduce the following generating functions characterizing the cavity population:

$$g_i(z_1, z_2, z_3, t)$$

8.1.2
$$= E\left[z_1^{X_1(t)}z_2^{X_2(t)}z_3^{X_3(t)}z^{W(t)} \mid X_i(0) = 1 = W(0), v = 0\right] \qquad i = 1, 2, 3$$

8.1.3
$$g(z_1, z_2, z_3, t) = E\left[z_1^{X_1(t)}z_2^{X_2(t)}z_3^{X_3(t)}z^{W(t)} \mid W(0) = 0, v \neq 0\right].$$

We next obtain a relation connecting g and g_1. We note that the conditioning on the r.h.s. of 8.1.3 implies that the population process is generated by immigrants; noting that the time to the arrival of the first immigrant is exponentially distributed with parameter v and that the immigrant will generate a population independent of further immigrants, we obtain the following equation:

$$g(z_1, z_2, z_3, z, t)$$

$$8.1.4 \qquad = e^{-vt} + v \int_0^t e^{-vu} g(z_1, z_2, z_3, z, t - u) g_1(z_1, z_2, z_3, z, t - u) \, du.$$

Solving the integral equation, we obtain

$$g(z_1, z_2, z_3, z, t)$$

$$8.1.5 \qquad = \exp - v \int_0^t [1 - g_1(z_1, z_2, z_3, z, u)] \, du.$$

The equilibrium distribution of the cavity is obtained by taking the limit as t goes to infinity.

8.1a   Evolution in phases

We next invoke the branching nature of the process to obtain an equation for g_1 and g_2. The constancy of the population rates implies Markov nature and hence by analyzing the various possibilities in the infinitesimal interval $(0, \Delta)$, we obtain the following backward differential relation:

$$g_1(z_1, z_2, z_3, z, t) = 1 - (\lambda + \mu + \eta) \Delta g_1(z_1, z_2, z_3, z, t - \Delta)$$

$$8.1.6 \qquad\qquad + \lambda \Delta g_2(z_1, z_2, z_3, z, t - \Delta) + (\mu + \eta) \Delta + o(\Delta)$$

by arguing that the photon that is in phase 1 initially can, in the interval $(0, \Delta)$,

(i) move to phase 2 with probability $\lambda \Delta + o(\Delta)$;

(ii) be absorbed or detected with probability $(\mu + \eta)\Delta + o(\Delta)$ and can continue to be in phase 1 with residual probability $1 - (\lambda + \mu + \eta)\Delta + o(\Delta)$.

Proceeding to the limit as $\Delta \to 0$, we obtain

$$\frac{\partial g_1(z_1, z_2, z_3, z, t)}{\partial t} = (\lambda + \mu + \eta) g_1(z_1, z_2, z_3, z, t)$$

$$8.1.7 \qquad\qquad + \lambda g_2(z_1, z_2, z_3, z, t) + \mu + \eta.$$

In an exactly similar manner, we obtain

$$\frac{\partial g_2(z_1, z_2, z_3, z, t)}{\partial t} = -(\lambda + \mu + \eta + \alpha)g_2(z_1, z_2, z_3, z, t)$$

$$+ \alpha g_2(z_1, z_2, z_3, z, t)g_1(z_1, z_2, z_3, z, t)$$

8.1.8
$$+ \lambda g_3(z_1, z_2, z_3, z, t) + \mu + \eta.$$

The product term $\alpha g_2 g_1$ arises because of the assumption that a photon in phase 2 can create a photon of age 0 (and hence in phase 1) at a constant rate α. The branching nature of the process leads to the product $g_1 g_2$.

In view of our assumption that the photons in phase 3 have zero birth rates, the function g_3 is independent of z_1 and z_2 and can be explicitly evaluated:

8.1.9
$$g_3(z_1, z_2, z_3, z, t) = 1 + (z_3 z - 1)e^{-(\mu + \eta)t}$$

It is rather difficult to solve for g_1 and g_2 explicitly from 8.1.7 and 8.1.8. However, the moments of $X_1(t)$ and $X_2(t)$ can be calculated.

8.1b Moments of the cavity population

We introduce the moments $a_k^i(t)$, $b_k^{ij}(t)$, $a^i(t)$, $b^{ij}(t)$ $(i, j = 1, 2, 3; k = 1, 2)$ by

8.1.10
$$a_k^i(t) = \left.\frac{\partial g_k}{\partial z_i}\right|_{z = z_1 = z_2 = z_3 = 1} \qquad a^i(t) = \left.\frac{\partial g}{\partial z_i}\right|_{z_1 = z_2 = z_3 = 1}$$

8.1.11
$$b_k^{ij}(t) = \left.\frac{\partial^2 g_k}{\partial z_i \partial z_j}\right|_{z_1 = z_2 = z_3 = 1} \qquad b^{ij}(t) = \left.\frac{\partial^2 g}{\partial z_i \partial z_j}\right|_{z = z_1 = z_2 = z_3 = 1}$$

We first connect the moments of $X_1(t)$ and $X_2(t)$ with the corresponding conditional moments by differentiating both sides of 8.1.5 appropriately:

8.1.12
$$a^i(t) = v \int_0^t a_1^i(u)\, du$$

8.1.13
$$b^{ij}(t) = v \int_0^t b_1^{ij}(u)\, du + a^i(t)a^j(t).$$

We next differentiate 8.1.7 and 8.1.8 to obtain

8.1.14
$$\frac{da_1^i(t)}{dt} + (\lambda + \mu + \eta)a_1^i(t) = \lambda a_2^i(t) \quad i = 1, 2, 3$$

8.1.15 $\qquad \dfrac{da_2^i(t)}{dt} + (\lambda + \mu + \eta)a_2^i(t) = \alpha a_1^i(t) \quad i = 1, 2$

8.1.16 $\qquad \dfrac{da_2^3(t)}{dt} + (\lambda + \mu + \eta)a_2^3(t) = \alpha a_1^3(t) + \lambda\, e^{-(\mu + \eta)t}.$

We note that the initial conditions are given by

8.1.17 $\qquad a_1^1(0) = a_2^2(0) = 1, \quad a_2^1(0) = a_1^2(0) = 0 = a_1^3(0) = a_2^3(0).$

Solving the system of linear differential equations 8.1.14 through 8.1.16, we obtain

8.1.18 $\qquad a_1^1(t) = a_2^2(t) = \tfrac{1}{2}\big[p(t) + q(t)\big]$

8.1.19 $\qquad a_2^1(t) = \dfrac{1}{2}\sqrt{\dfrac{\alpha}{\lambda}}\,\big[p(t) - q(t)\big]$

8.1.20 $\qquad a_1^2(t) = \dfrac{1}{2}\sqrt{\dfrac{\lambda}{\alpha}}\,\big[p(t) - q(t)\big]$

8.1.21 $\qquad a_1^3(t) = \dfrac{\lambda}{\lambda - \alpha}\,\{e^{-(\mu + \eta)t} - a_1^2(t) - a_1^1(t)\}$

8.1.22 $\qquad a_2^3(t) = a_1^2(t) + a_1^3(t)$

where $p(t)$ and $q(t)$ are given by

8.1.23 $\qquad p(t) = \exp\,-(\lambda + \mu + \eta - \sqrt{\lambda\alpha t})$

8.1.24 $\qquad q(t) = \exp\,-(\lambda + \mu + \eta + \sqrt{\lambda\alpha t}).$

Again differentiating 8.1.7 and 8.1.8 successively, we obtain for $i, j = 1, 2, 3$

8.1.25 $\qquad \dfrac{db_1^{ij}(t)}{dt} + (\lambda + \mu + \eta)b_1^{ij}(t) = \lambda b_2^{ij}(t)$

8.1.26 $\qquad \dfrac{db_2^{ij}(t)}{dt} + (\lambda + \mu + \eta)b_2^{ij}(t) = \alpha b_1^{ij}(t) + \alpha\big[a_1^i(t)a_2^j(t) + a_1^j(t)a_2^i(t)\big].$

The system of equations 8.1.26 and 8.1.27 can be solved to arrive at explicit expressions for $b_k^{ij}(t)$. However, we do not need them; we shall obtain the LT of $b_k^{ij}(t)$ from which the relevant information regarding the detection process can be extracted. Denoting by $b_k^{ij}{*}(s)$ the LT of $b_k^{ij}(t)$, we obtain for $i, j = 1, 2, 3$

8.1.27 $\qquad b_1^{ij}{*}(s) = \lambda\alpha L^{ij}{*}(s)/D(s)$

8.1.28 $\qquad b_2^{ij}{*}(s) = (\lambda + \mu + s)b_1^{ij}(s)/\lambda,$

where

8.1.29 $\qquad D(s) = (s + \lambda + \eta + \mu)^2 - \lambda\alpha$

8.1.30 $\qquad 2\lambda L^{11*}(s) = 2\alpha L^{22*}(s) = \lambda\alpha/D(s/2)$

8.1.31 $\qquad 4L^{21*}(s) = 4L^{12*}(s) = (s + 2\lambda + 2\mu + 2\eta)/D(s/2)$

8.1.32 $\qquad L^{31*}(s) = L^{13*}(s) = \dfrac{\lambda}{\lambda - \alpha}\{a_1^{1*}(s + \mu + \eta) + a_2^{1*}(s + \mu + \eta) - $

$$-\tfrac{1}{2}L^{11*}(s) - \tfrac{1}{2}L^{22*}(s) - L^{12*}(s)\}$$

8.1.33 $\qquad L^{23*}(s) = L^{32*}(s) = \dfrac{\lambda}{\lambda - \alpha}\{a_1^{1*}(s + \mu + \eta) + a_1^{2*}(s + \mu + \eta) - $

$$- L^{22*}(s) - L^{12*}(s)\}$$

$$L^{33*}(s) = \left(\frac{\lambda}{\lambda - \alpha}\right)^2\left\{\frac{1}{s + 2\mu + 2\eta} - 2a_1^{1*}(\mu + \eta + s) - \right.$$

8.1.34 $$\qquad -\left(\frac{\lambda + \alpha}{\lambda}\right)a_1^{2*}(\mu + \eta + s) + \left(\frac{\lambda + \alpha}{2\lambda}\right)L^{22*}(s) +$$

$$+ \tfrac{1}{2}[a_1^{1*}(s + \mu + \lambda + \eta - \sqrt{\lambda\alpha}) + a_1^{1*}(s + \mu + \lambda + \eta + \sqrt{\lambda\alpha})] +$$

$$\left. + \frac{1}{2}\sqrt{\frac{\lambda}{\alpha}}[a_1^{2*}(s + \mu + \lambda + \eta - \sqrt{\lambda\alpha}) - a_1^{2*}(s + \mu + \lambda + \eta + \sqrt{\lambda\alpha})]\right\}.$$

The moments and cross-correlations of the equilibrium distribution can be obtained from 8.1.12 and 8.1.13 by proceeding to the limit as $t \to \infty$. Thus we have

$$a^1(\infty) = v(\lambda + \mu + \eta)/D(0),$$

$$a^2(\infty) = v\lambda/D(0)$$

8.1.35 $\qquad a^3(\infty) = v\lambda^2/[(\mu + \eta)D(0)]$

$$b^{11}(\infty) = v\lambda\alpha^2/[2[D(0)]^2] + [a^1(\infty)]^2$$

$$b^{22}(\infty) = v\lambda^2\alpha/[2[D(0)]^2] + /a^2(\infty)]^2$$

$$b^{21}(\infty) = b^{12}(\infty) = v\lambda(\lambda + \mu + \eta)\alpha/[2[D(0)]^2] + a^1(\infty)a^2(\infty).$$

Expressions for the cross-correlations of the type b^{i3} do not take a simple form; nevertheless they can be evaluated by the combined use of 8.1.13 and 8.1.32 through 8.1.34.

We obtain an interesting result if we neglect the cavity photons in phase 3. Let $W(t)$ represent the total number of cavity photons in phases 1 and 2. Then from 8.1.35 we have

$$8.1.36 \qquad E[W(\infty)] = v(2\lambda + \mu + \eta)/D(0)$$

$$E\{W(\infty)[W(\infty) - 1]\}$$

$$8.1.37 \qquad = (E[W(\infty)])^2\left\{1 + \frac{\lambda\alpha(3\lambda + 2\mu + 2\eta + \alpha)}{2v(2\lambda + \mu + \eta)^2}\right\}.$$

The expression within braces on the r.h.s. of 8.1.37 represents a measure of bunching of the cavity photon population. If we choose λ very large, bunching can be considerably reduced by choosing α small. If we set $\alpha = \lambda$, it reduces to $1 + \lambda^2/[v(2\lambda + \mu + \eta)]$. The persistence of bunching is to be expected from a model of population growth. The motivation to set $\alpha = \lambda$ will become clear in the following section; however, this makes the model rather perplexing, for if we choose $v = \lambda^2/(2\lambda + \mu + \eta)$, we can recover the result corresponding to Bose–Einstein distribution. An interesting feature is that v must be chosen smaller (than say λ) in order to achieve a bunching factor of 2.

8.1c   Behaviour of the third moment

We pursue further the special case when $\lambda = \alpha$ and $v = \lambda^2/(2\lambda + \mu + \eta)$. We proceed to obtain the third factorial moment of the equilibrium population. Introducing the notation

$$8.1.38 \qquad c_i(t) = \frac{\partial^3 g_i}{\partial z^3}(z_1, z_2, z_3, z, t)\big|_{z_1 = z_2 = z_3 = z = 1}$$

we obtain

$$8.1.39 \qquad \frac{dc_1(t)}{dt} = -(\lambda + \mu + \eta)c_1(t) + \lambda c_2(t)$$

$$8.1.40 \qquad \frac{dc_2(t)}{dt} = -(\lambda + \mu + \eta)c_2(t) + \lambda c_1(t) + 3\lambda(b_2(t)a_1(t) + a_2(t)b_1(t))$$

with initial conditions

$$8.1.41 \qquad c_i(0) = 0.$$

In our case $a_i(t)$ and $b_i(t)$ are given by

8.1.42 $\qquad a_1(t) = a_2(t) = e^{-(\mu + \eta)t}$

8.1.43 $\qquad b_1(t) + b_2(t) = 2\lambda e^{-(\mu + \eta)t}[1 - e^{-(\mu + \eta)t}]/(\mu + \eta).$

Solving 8.1.39 and 8.1.40 by Laplace transform, we obtain

8.1.44 $\qquad c_1^*(s) = 3\lambda^2 L(s)/[(\lambda + \mu + \eta + s)^2 - \lambda^2]$

where $L(s)$ is given by

$$8.1.45 \qquad L(s) = \int_0^\infty a_1(t)(b_2(t) + b_1(t)) e^{-st}\, dt$$

$$= 2\lambda/(s + 2\mu + 2\eta)(s + 3\mu + 3\eta).$$

Thus we finally have

8.1.46 $\qquad c_1^*(0) = \lambda^3/(\eta + \mu)^3(2\lambda + \mu + \eta).$

The third factorial moment of the equilibrium population of cavity photons is given by

$$8.1.47 \qquad \lim_{t \to \infty} C(t) = \lim_{t \to \infty} \frac{\partial^3 g}{\partial z^3}(z_1, z_2, z_3, z, t)\big|_{z_1 = z_2 = z_3 = z = 1};$$

using 8.1.4, we finally obtain

$$8.1.48 \qquad C(\infty) = [va_1^*(0)]^3 + 3v^2 a_1^*(0)b_1^*(0) + vC_1^*(0)$$

$$= \left(\frac{v}{\mu + \eta}\right)^3 \left(6 + \frac{\mu + \eta}{\lambda}\right).$$

It is interesting to note that for thermal light we should have

$$8.1.49 \qquad C(\infty) = 6\left(\frac{v}{\mu + \eta}\right)^3.$$

Thus if we choose $(\mu + \eta)/\lambda$ small, then the resulting radiation will correspond to Gaussian light. For instance, if $(\mu + \eta)/\lambda = 1/16$, the discrepancy is only one percent.

8.1d Arbitrary number of phases

Extension of the analysis to general age-dependent birth rate of the form 8.1.1, where $n\,(>2)$ is any arbitrary positive integer, is straightforward. The life span of any photon (conditional upon its survival over the span) can be thought of as the sum of spans of $(n + 1)$ phases. The spans of the first n

phases are independent and exponentially distributed with the same parameter λ; the $(n+1)$th phase is defined as the residue. We assume further that the birth rate is zero in each of the first $(n-1)$ phases, while it is a constant equal to α during the nth phase; the birth rate is again assumed to be zero in the residuary $(n+1)$th phase. The probability that a photon of specified age x is in phase k is simply

$$e^{-\lambda x}(\lambda x)^k/k! \quad (k \geqslant 1)$$

so that the probability that a photon of given age x at time t produces another (of age zero) in the interval $(t, t+dt)$ is simply $\lambda(x)\,dt + o(dt)$ where $\lambda(x)$ is given by 8.1.1. If we denote by $X_i(t)$ the random variable representing the number of photons in phase i at time t, and by $X(t)$ the total number of photons at time t, we can introduce the generating function $g_j(z_1, z_2, \ldots, z_n, t)$ $(j = 1, 2, \ldots, n)$ by

$$g_j(z_1, z_2, \ldots, z_n, t)$$

8.1.50
$$= E\left[\prod_{i=1}^{n} z_i^{X_i(t)}\right) z^{X(t)} \mid X_j(0) = X(0) = 1, v = 0\right].$$

We find that the backward differential relation leads to

8.1.51
$$\frac{\partial g_i}{\partial t} = -(\lambda + \mu + \eta)g_i + \lambda g_{i+1} + \mu + \eta \quad (1 \leqslant i < n)$$

8.1.52
$$\frac{\partial g_n}{\partial t} = -(\lambda + \mu + \eta + \alpha)g_n + \alpha g_n g_1 + \lambda + \mu + \eta$$

with the initial conditions

8.1.53
$$g_i(z_1, z_2, \ldots, z_n, z, 0) = z_i z.$$

The analysis leading to the determination of the moments and cross-correlations can be carried through as in section 8.1. The first moments of $X_i(t)$ can be explicitly obtained as weighted negative exponentials under $(\mu + \eta - \lambda - \Gamma \omega)t$, where $\Gamma^n = \lambda^{n-1}\alpha$ and ω is an nth root of unity. The second-order moments and correlations can also be obtained; at any rate the LT solutions of the moments can be obtained explicitly. The analysis of the detection process will also follow on lines very similar to those presented earlier in this section.

8.2 Modified Kendall evolution: detection process

We are generally interested in $N(t_0, t)$, the number of photons detected in a destructive scheme of counting over the interval $(t_0, t_0 + t)$. In the model under consideration, the process is stationary and hence the distributional characteristics of $N(t_0, t)$ are independent of t_0. The physically important case corresponds to the situation when the cavity is maintained in a state of equilibrium say at the time-point t_0. Now there are two ways of dealing with the detection process. The first is to deal directly with the probability distribution of the number of photocounts over the interval $(t_0, t_0 + t)$; the second method consists in dealing with the point process generated by the epochs of detection. The probability distribution of $N(t_0, t)$, or rather of $N(t)$, is best studied by introducing a more comprehensive generating function $G_1(z_1, z_2, z_3, z, t)$, where

$$G_i(z_1, z_2, z_3, z, t)$$

8.2.1
$$= E\left[z_1^{X_1(t)} z_2^{X_2(t)} z_3^{X_3(t)} z^{N(t)} \mid X_i(0) = 1 = W(0), v = 0\right] \quad (i = 1, 2, 3)$$

8.2.2
$$G(z_1, z_2, z_3, z, t) = E\left[z_1^{X_1(t)} z_2^{X_2(t)} z_3^{X_3(t)} z^{N(t)} \mid W(0) = 0, v \neq 0\right].$$

Now the photocount distribution corresponding to equilibrium initial condition can be obtained from the generating function $G_E(z, t)$ defined by

8.2.3
$$G_E(z, t) = E\left[z^{N(t)} \mid \text{cavity population in eqm at the time-origin}\right]$$

The equilibrium distribution of the cavity population can be obtained by setting $z = 1$ and taking the limit as $t \to \infty$ in 8.2.2. If we use the notation

8.2.4
$$G_i(z, t) = G_i(1, 1, 1, z, t)$$

8.2.5
$$P_E(z_1, z_2, z_3) = G(z_1, z_2, z_3, 1, \infty)$$

we can use the independence of the population evolution process to obtain

8.2.6
$$G_E(z, t) = G(z, t)P_E(G_1(z, t), G_2(z, t), G_3(z, t)).$$

We proceed as in 8.1 to obtain backward differential equations satisfied by the generating functions. Thus we obtain

8.2.7
$$\frac{\partial G_1(z_1, z_2, z_3, z, t)}{\partial t} = -(\lambda + \mu + \eta)G_1(z_1, z_2, z_3, z, t)$$
$$+ \lambda G_2(z_1, z_2, z_3, z, t) + \mu + \eta z$$

8.2.8
$$\frac{\partial G_2(z_1, z_2, z_3, z, t)}{\partial t} = -(\lambda + \mu + \eta + \alpha)G_2(z_1, z_2, z_3, z, t)$$
$$+ \alpha G_2(z_1, z_2, z_3, z, t) + G_1(z_1, z_2, z_3, z, t)$$
$$+ \lambda G_3(z_1, z_2, z_3, z, t) + \mu + \eta z$$

8.2.9
$$\frac{\partial G(z_1, z_2, z_3, z, t)}{\partial t} = -\nu G(z_1, z_2, z_3, z, t) + \nu G(z_1, z_2, z_3, z, t)$$
$$\times G_1(z_1, z_2, z_3, z, t)$$

with the initial conditions

8.2.10
$$G_i(z_1, z_2, z_3, z, t) = z_i \quad (i = 1, 2)$$

8.2.11
$$G(z_1, z_2, z_3, z, t) = 1.$$

The function G_3 can be explicitly evaluated as before:

8.2.12
$$G_3(z_1, z_2, z_3, z, t) = z_3 e^{-(\mu + \eta)t} + \frac{\mu + \eta z}{\mu + \eta}(1 - e^{-(\mu + \eta)t}).$$

The moments of the photocount distribution can be obtained by appropriate differentiation of equations 8.2.7 through 8.2.9 and using the connecting relations 8.2.4, 8.2.5 and 8.2.6. We do not propose to go further in this direction particularly in view of the results that are already available in the previous section. The point process of detection can be characterized in terms of the sequence of conditional product densities defined by

8.2.13
$$f_1(t) = \lim_{\Delta \to 0} Pr\{N(t + \Delta) - N(t) = 1 \,|\, \text{cavity in eqm initially}\}/\Delta$$

$$f_2(t_1, t_2) = \lim_{\Delta_1, \Delta_2 \to 0} Pr\{N(t_1 + \Delta_1) - N(t_1) = 1, N(t_2 + \Delta_2) -$$

8.2.14
$$- N(t_2) = 1 \,|\, \text{cavity in eqm initially}\}/\Delta_1 \Delta_2.$$

It is clear from the construction of our model that the process is stationary, and hence we have

8.2.15
$$f_1(t) = \text{a constant}$$

8.2.16
$$f_2(t_1, t_2) = \text{a function of } |t_2 - t_1| = h_{sty}(|t_2 - t_1|).$$

Our primary object is to identify the constant on the r.h.s. of 8.2.15 and obtain the function $h_{sty}(.)$ explicitly.

We proceed as before, to obtain these functions by appropriate

154

conditioning at the origin and then revert to the equilibrium condition. Accordingly we define

$$_ih_1^i(t) = \lim_{\Delta \to 0} Pr\{N(t + \Delta) - N(t) = 1 \mid X_1(0) = 2 - i,$$

8.2.17
$$X_2(0) = i - 1, v = 0\} \quad (i = 1, 2)$$

8.2.18
$$h_1(t) = \lim_{\Delta \to 0} Pr\{N(t + \Delta) - N(t) = 1 \mid X(0) = 0 = Y(0), v \neq 0\}/\Delta.$$

We note that a photon in phase 3 is subject to absorption or detection and hence it can be dealt with directly. To start with, we shall assume that photons in all the phases are subject to detection.

Next we note that $h_1(.)$ and $f_1(.)$ are connected by

8.2.19
$$f_1(\cdot) = \lim_{t \to \infty} h_1(t).$$

The Poisson nature of the immigration process implies

8.2.20
$$h_1(t) = v \int_0^t e^{-v\tau}[h_1(t - \tau) + {}_ih_1^1(t - \tau)]\,d\tau.$$

The above relation is obtained by arguing that the event corresponding to detection at epoch t may be generated by the first immigration photon or by a subsequent one. Solving the above relation, we obtain

8.2.21
$$h_1(t) = v \int_0^t {}_ih_1^1(\tau)\,d\tau.$$

Using the definition of $_ih_1^1(.)$ we obtain directly

8.2.22
$$_ih_1^1(t) = \eta[a_1^1(t) + a_1^2(t) + a_1^3(t)],$$

where $a_i^j(.)$ are the conditional first moments introduced in 8.1 and explicitly given by 8.1.18 through 8.1.20. Using those results, we obtain

8.2.23
$$f_1(.) = \frac{v\eta}{\mu + \eta}\left[1 + \frac{\alpha\lambda}{(\lambda + \mu + \eta)^2 - \alpha\lambda}\right],$$

where the first term on the r.h.s. corresponds to the contribution from the third phase.

In order to obtain an expression for $h_{sty}(t)$ $(t > 0)$ we note that for the two detections separated by t, the contribution can arise from the same population tree or from different trees. Using combinatorial arguments, as in section 7.6, we have

8.2.24 $\qquad h_{sty}(t) = f_1(\,.\,)h_1(t) + \eta \sum_{i,j} b^{ij}{}_I h^j_i(t),$

where the b^{ij} are the equilibrium second moments of the cavity population and the summation over i,j runs from 1 to 3. The function $h^2_1(\,.\,)$ is obtained on the same lines as before:

8.2.25 $\qquad h^2_1(t) = \eta[a^1_2(t) + a^2_2(t) + a^3_2(t)].$

On the other hand, the function $h^3_1(\,.\,)$ is obtained directly by observing that the initial photon in phase 3 must survive absorption and detection right up to t:

8.2.26 $\qquad {}_I h^3_1(t) = \eta\,e^{-(\mu + \eta)t}.$

The constants b^{ij} are obtained by taking the limit as $t \to \infty$ of $b^{ij}(t)$ defined by 8.1.13.

The measure $\mathscr{B}$ of bunching of the point events of the detection process defined by

8.2.27 $\qquad \mathscr{B} = h_{sty}(0)/h_{sty}(\infty)$

can be obtained by observing that

8.2.28 $\qquad h_0(0) = 0, \quad {}_I h^i_1(\infty) = 0, \quad {}_I h^i_1(0) = \eta:$

8.2.29 $\qquad \mathscr{B} = \sum_{i,j=1}^{3} b^{ij}/[f_1(\,.\,)]^2.$

Using the relation 8.1.35, we can conclude that $\mathscr{B} > 1$.

Next we consider the situation where the population in phase 3 is lost; this can be interpreted as an asymmetric behaviour of the phases, the asymmetry arising from a higher rate of absorption equal to $\mu + \lambda$ in the second phase. We further set $\lambda = \alpha$; then 8.2.24 can be cast in the form

$$h_{sty}(t) = h_1(\infty)h_1(t)$$

8.2.30 $\qquad\qquad + \eta[(b^{11} + b^{21})_I h^1_1(t) + (b^{22} + b^{12})_I h^2_1(t)],$

where

8.2.31 $\qquad h_1(\infty) = v\eta(2\lambda + \mu + \eta)/D(0)$

8.2.32 $\qquad {}_I h^1_1(t) = {}_I h^2_1(t) = \eta\,e^{-(\mu + \eta)t}.$

Substituting the values for b^{ij} from 8.1.35, we finally arrive at the following simple formula:

$$8.2.33 \qquad h_{sty}(t) = \frac{v\eta^2}{(\mu+\eta)^2} \left\{ v + \frac{\lambda^2 e^{-(\mu+\eta)t}}{2\lambda+\mu+\eta} \right\}.$$

Then if we make the choice

$$8.2.34 \qquad v = \frac{\lambda^2}{2\lambda+\mu+\eta},$$

we arrive at

$$8.2.35 \qquad h_{sty}(t) = (h(\infty))^2 (1 + e^{-(\mu+\eta)t})$$

a formula corresponding to thermal light with a Lorentzian spectrum. This is an interesting result. The second-order properties of the detection process arising from an age-dependent (and hence non-Markov) population point process coincide with those of the detection process arising from a population point process with constant rates. The choice $\lambda = \alpha$ makes the process even, in the sense that the rate of production equals the rate of loss to the third phase. However, the choice of v does not admit of a direct interpretation. Further research is necessary to settle the question whether the resulting detection process can be made identical with the photodetection process of thermal light.

Finally we retain the choice $\lambda = \alpha$ but take into account the photons in phase 3. In this case we have

$$8.2.36 \qquad a_1^3(t) = \frac{e^{-(\mu+\eta)t}}{4} \left[e^{-2\lambda t} + 2\lambda t - 1 \right]$$

and consequently $h_1^1(t)$ and $h_1^2(t)$ are given by

$$8.2.37 \qquad {}_1 h_1^1(t) = \frac{e^{-(\mu+\eta)t}}{4} \eta (e^{-2\lambda t} + 2\lambda t + 3)$$

$$8.2.38 \qquad {}_1 h_1^2(t) = \frac{e^{-(\mu+\eta)t}}{4} \eta (5 - 2e^{-2\lambda t} + 2\lambda t).$$

Thus $h_1(t)$ takes the form

$$8.2.39 \qquad h_1(t) = \frac{\eta}{4} \left\{ \frac{1}{\mu+\eta+2\lambda} + \frac{3(\mu+\eta)+2\lambda}{(\mu+\eta)^2} - \frac{e^{-(\mu+\eta+2\lambda)t}}{\mu+\eta+2\lambda} \right.$$
$$\left. - \frac{(\mu+\eta)(1+2\lambda t) + 2\lambda e^{-(\mu+\eta)t}}{(\mu+\eta)^2} \right\}.$$

We can substitute 8.2.37 through 8.2.39 in 8.2.24 to arrive at the final formula for $h_{sty}(t)$. It is to be specially noted that the spectrum corresponding to $h_{sty}(t)$ contains new features characteristic of non-Markov evolution. In particular the spectrum shows the existence of terms corresponding to Poisson profiles. Such spectra were generated earlier in the context of Gaussian light by Srinivasan and Sukavanam (1971, 1972).

8.3 General phase-model: evolution through four phases

We next analyze a more complex type of non-Markov evolution by generalizing the phase-model corresponding to the age-dependent birth rate 8.1.1 when $n = 3$. There are four phases; the life-spans of the first three phases are assumed to be exponentially distributed with parameters λ_1, λ_2, and λ_3. Moreover the death (absorption) rates in the first three phases are assumed to be constants, respectively equal to μ_1, μ_2 and μ_3. In the last phase the death rate is assumed to be the same as in the third phase. The birth rate is zero in all phases except the third, where it takes the value α. As usual, we assume that spontaneous emissions are at a constant rate v so that they form a Poisson process with parameter v. The detection is assumed to be a constant rate per unit time. The primary motivation for the use of differential death rates stems from the fact that absorption is also part of evolution and hence the non-Markov nature should get reflected in it too.

8.3a Generating functions

We introduce the following notation:

$X_i(t)$: number of photons in phase i at time t;

$X(t)$: total number of photons at time t;

$Y(t)$ number of photons detected over the time-interval $(0, t)$.

A comprehensive description of the population can be provided by introducing the generating functions g_i $(i = 1, 2, 3, 4)$ and g by

$$g_i(z_1, z_2, z_3, z, w, t)$$

8.3.1
$$= \left[z_1^{X_1(t)} z_2^{X_2(t)} z_3^{X_3(t)} z^{X(t)} w^{Y(t)} \,\big|\, X_i(0) = X(0) = 1, v = 0 \right]$$

$$g(z_1, z_2, z_3, z, w, t)$$

8.3.2
$$= E\left[z_1^{X_1(t)} z_2^{X_2(t)} z_3^{X_3(t)} z^{X(t)} w^{Y(t)} \,\big|\, X(0) = 0, v \neq 0 \right].$$

The connecting relation between g and g_1 remains the same as before:

$$8.3.3 \qquad g(z_1, z_2, z_3, z, w, t) = \exp - v \int_0^t \left[1 - g_1(z_1, z_2, z_3, z, w, \tau) \right] d\tau$$

and g_4 is given by

$$8.3.4 \qquad g_4(z_1, z_2, z_3, z, w, t) = z\,e^{-(\mu+\eta)t} + \frac{\mu + \eta w}{\mu + \eta} \left(1 - e^{-(\mu+\eta)t} \right).$$

We shall keep out of our purview photons in phase 3; alternatively we can assume the death rate in phase 3 to be $\lambda_3 + \mu_3$.

As before, we can use the backward differential relation to obtain

$$8.3.5 \qquad \frac{\partial g_i}{\partial t} = -(\lambda_i + \mu_1 + \eta)g_i + \lambda_i g_{i+1} + \mu_i + \eta z \quad i = 1, 2$$

$$8.3.6 \qquad \frac{\partial g_3}{\partial t} = -(\lambda_3 + \mu_3 + \eta + \alpha)g_3 + \alpha g_3 g_1 + \lambda_3 + \mu_3 + \eta z$$

with the initial conditions

$$8.3.7 \qquad g_i(z_1, z_2, z_3, z, w, 0) = z_i z.$$

8.3b   Factorial moments and cross-correlations

We introduce the factorial moments and cross-correlations by

$$a_i^j(t) = \frac{\partial g_i}{\partial z^j}, \quad a_i(t) = \frac{\partial g_i}{\partial z}$$

$$b_i^{jk}(t) = \frac{\partial^2 g_i}{\partial z_j \partial z_k}$$

$$8.3.8 \qquad a^j(t) = \frac{\partial g}{\partial z_j}, \quad b^{jk}(t) = \frac{\partial^2 g}{\partial z_j \partial z_k},$$

where the derivatives are evaluated at the point $z_1 = z_2 = z_3 = w = z = 1$. We first note

$$8.3.9 \qquad a_i(t) = \sum_j a_i^j(t).$$

By differentiation of both sides of 8.3.5 and 8.3.6 we obtain

$$8.3.10 \qquad \frac{dz_1^j(t)}{dt} = -\alpha_1 a_1^j(t) + \lambda_1 a_2^j(t)$$

$$8.3.11 \qquad \frac{da_2^j(t)}{dt} = -\alpha_2 a_2^j(t) + \lambda_2 a_3^j(t)$$

$$8.3.12 \qquad \frac{da_3^j(t)}{dt} = -\alpha_3 a_3^j(t) + \alpha a_1^j(t)$$

with the initial conditions $a_1^j(0) = \delta_{ij}$; the constants α_i are given by

$$8.3.13 \qquad \alpha_i = \lambda_i + \mu_i + \eta \quad i = 1, 2, 3.$$

We solve the above set of equations by Laplace transforms; thus we obtain after some computations,

$$8.3.14 \qquad a_i^{j*}(s) = F_i^j(s)/D(s), \quad a_i^*(s) = F_i(s)/D(s)$$

where

$$8.3.15 \qquad D(s) = (\alpha_1 + s)(\alpha_2 + s)(\alpha_3 + s) - \lambda_1 \lambda_2 \alpha$$

$$F_1^1(s) = (\alpha_2 + s)(\alpha_3 + s), \quad F_1^2(s) = \lambda_1(\alpha_3 + s), \quad F_1^3(s) = \lambda_1 \lambda_2$$

$$F_2^1(s) = \alpha \lambda_2, \quad F_2^2(s) = (\alpha_1 + s)(\alpha_3 + s), \quad F_2^3(s) = \lambda_2(\alpha_1 + s)$$

$$F_3^1(s) = \alpha(\alpha_2 + s), \quad F_3^2(s) = \alpha \lambda_1, \quad F_3^3(s) = (\alpha_1 + s)(\alpha_2 + s)$$

$$F_1(s) = (\alpha_3 + s)(\alpha_2 + s + \lambda_1), \quad F_2(s) = (\alpha_1 + s)(\alpha_3 + s + \lambda_2) + \alpha \lambda_2$$

$$8.3.16 \qquad F_3(s) = (\alpha_2 + s)(\alpha_1 + s + \alpha) + \alpha \lambda_1.$$

Let $\Gamma_1, \Gamma_2, \Gamma_3$ be the roots of the equation $D(s) = 0$; then the solution can be written as

$$a_i^j(t) = \frac{F_i^j(\Gamma_1) e^{\Gamma_1 t}}{(\Gamma_1 - \Gamma_2)(\Gamma_1 - \Gamma_3)} + \frac{F_i^j(\Gamma_2) e^{\Gamma_2 t}}{(\Gamma_2 - \Gamma_1)(\Gamma_2 - \Gamma_3)}$$

$$8.3.17 \qquad + \frac{F_i^j(\Gamma_3) e^{\Gamma_3 t}}{(\Gamma_3 - \Gamma_1)(\Gamma_3 - \Gamma_2)} \qquad (i, j = 1, 2, 3).$$

We can follow exactly the method of analysis presented in section 8.2 and obtain explicit expressions for the LT of the b-functions. However, we will not do so; rather we will consider a special case that will turn out to be interesting. We put the following condition on the zeros Γ_i of $D(s)$:

$$8.3.18 \qquad 2\Gamma_3 = \Gamma_1 + \Gamma_2.$$

The condition will prepare the ground for identifying the radiation as the one corresponding to the superposition of the Lorentzian profiles. The condition 8.3.18 will be satisfied if the parameters α_i, λ_i, and α satisfy the condition

$$(\alpha_1 + \alpha_2 + \alpha_3)\left[-\tfrac{5}{18}(\alpha_1 + \alpha_2 + \alpha_3)^2 + \tfrac{1}{2}(\alpha_1^2 + \alpha_2^2 + \alpha_3^2)\right]$$

8.3.19
$$= 3(\alpha\lambda_1\lambda_2 - \alpha_1\alpha_2\alpha_3).$$

It can be verified that 8.3.19 is a feasible condition in the sense that physically meaningful values α_i, λ_i and α satisfying 8.3.19 do exist. Using 8.3.18, we can rewrite 8.3.17 in the form

$$a_i^j(t) = 2\{F_i^j(\Gamma_1)\,e^{\Gamma_1 t} + F_i^j(\Gamma_2)\,e^{\Gamma_2 t}$$

8.3.20
$$- 2F_i^j(\Gamma_3)^{\Gamma_3 t}\}/(\Gamma_1 - \Gamma_2)^2 \quad (i,j = 1,2,3).$$

It then follows that

$$a_i(t) = 2\{F_i(\Gamma_1)\,e^{\Gamma_1 t} + F_i(\Gamma_2)\,e^{\Gamma_2 t}$$

8.3.21
$$- 2F_i(\Gamma_3)\,e^{\Gamma_3 t}\}/(\Gamma_1 - \Gamma_2)^2 \quad (i = 1,2,3).$$

We next perform appropriate differentiations to obtain for $i,j = 1,2,3$

8.3.22
$$\frac{db_1^{ij}(t)}{dt} = -\alpha_1 b_1^{ij}(t) + \lambda_1 b_2^{ij}(t)$$

8.3.23
$$\frac{db_2^{ij}(t)}{dt} = -\alpha_2 b_2^{ij}(t) + \lambda_2 b_3^{ij}(t)$$

$$\frac{db_3^{ij}(t)}{dt} = -\mu_3 b_3^{ij}(t) + \alpha b_1^{ij}(t)$$

8.3.24
$$+ \alpha[a_3^j(t)a_1^i(t) + a_3^i(t)a_1^j(t)]$$

with the initial condition

8.3.25
$$b_k^{ij}(0) = 0 \qquad k = 1,2,3.$$

Using Laplace transform technique, we obtain

8.3.26
$$b_1^{ij*}(s) = \alpha\lambda_1\lambda_2 \int_0^\infty [a_3^j(t)a_1^i(t) + a_3^i(t)a_1^j(t)]\,e^{-st}\,dt/D(s).$$

For the special case when 8.3.18 holds, further simplification can be effected, leading to

$$b_1^{ij*}(0) = 2\alpha\lambda_1\lambda_2 \sum_k \xi_k[a_3^{j*}(-\Gamma_k)F_1^i(\Gamma_k)$$

8.3.27
$$+ a_3^{i*}(-\Gamma_k)F_1^j(\Gamma_k)]/[(\Gamma_1 - \Gamma_2)^2 D(0)]$$

where

8.3.28 $\qquad \xi_k = 1$ for $k = 1, 2;$ $\qquad \xi_3 = -2.$

Using 8.3.3, we finally obtain

8.3.29 $\qquad b^{ij}(\infty) = v^2 a_1^{i*}(0) a_1^{j*}(0) + v b_1^{ij*}(0),$

so that 8.3.29 taken with 8.3.27 yields an explicit formula for the second factorial and cross-moments of the stationary cavity population in different phases.

8.3c   A special model: superposition of two fields

The product densities of the detection can be obtained in exactly the same manner as in section 8.2. In this case we find that when the cavity and detector are conditioned to be in equilibrium the functions $f_1(.)$ and $h_{sty}(.)$ as defined by 8.2.15 and 8.2.16 are given by

8.3.30 $\qquad f_1(.) = h_1(\infty) = \eta v a_1^*(0)$

$$= \eta v[\alpha_2 \alpha_3 + \lambda_1 \alpha_3 + \lambda_1 \lambda_2]/D(0),$$

8.3.31 $\qquad h_{sty}(t) = (h_1(\infty))^2 + 2\eta^2 \sum E_i \xi_i \, e^{\Gamma_i t}/(\Gamma_1 - \Gamma_2)^2$

where

8.3.32 $\qquad E_i = h_1(\infty) v F_1(\Gamma_i)/(\eta \Gamma_i) + \sum_{jk} b^{jk} F_k(\Gamma_i).$

Next we observe (see Chapter 6) that the radiation resulting from the four-phase model has thermal characteristics if

8.3.33 $\qquad 2[h_1(\infty)]^2 = 2\eta^2 \sum_{ijk} \xi_i b^{jk} F_k(\Gamma_i)/(\Gamma_1 - \Gamma_2)^2.$

In view of 8.3.30, we have the identity

8.3.34 $\qquad 2\sum_i \xi_i F_k(\Gamma_i) = (\Gamma_1 - \Gamma_2)^2$

which reduces 8.3.33 to

8.3.35 $\qquad 2(h_1(\infty))^2 = \eta^2 \sum_{jk} b^{jk},$

a relation which expresses the thermal condition directly in terms of the moments.

We establish the feasibility of 8.3.33 and 8.3.19 by taking some specific values for the parameters; this was done by a trial-and-error method.

Actually there are too many parameters at our disposal; we fix some of them by making the choice

8.3.36 $\alpha_1 = 7\mu, \quad \alpha_2 = 3\mu, \quad \alpha_3 = 2\mu,$

where μ is some parameter having the dimension of (1/time). The condition 8.3.19 now reduces to

8.3.37 $\alpha\lambda_1\lambda_2 = 6\mu^3,$

while the Γ_i themselves are given by

8.3.38 $\Gamma_1 = -1.354\mu, \quad \Gamma_2 = -6.646\mu, \quad \Gamma_3 = -4\mu.$

There is a choice of λ_1, λ_2 and α, given by

8.3.39 $\lambda_1 = 4.62\mu, \quad \lambda_2 = \mu, \quad \alpha = 1.299\mu,$

satisfying 8.3.37 which when substituted in 8.3.35 determines v:

8.3.40 $v = 0.644\mu.$

For this choice of the parameters, we have

8.3.41 $f_1(.) = h_1(\infty) = 0.3553\eta$

8.3.42 $h_{sty}(t) = \eta^2(0.3553)^2 + 0.1165(e^{-0.677\mu|t|} + 0.0402\, e^{-3.223\mu|t|})^2\eta^2.$

If at this stage we compare the formula with the one in section 5.3 or 5.4, we can identify the correlation. For Gaussian light, we have the identification

8.3.43 $h_{sty}(t) = \bar{I}^2 + |G^{(1)}(rt_1, rt_2)|^2,$

where

8.3.44 $G^{(1)}(rt_1, rt_2) = (\bar{I}/1.0402)(e^{-0.677\mu|t|} + 0.402e^{-3.223\mu|t|}),$

leading to the conclusion that the resulting radiation corresponds to the superposition of two independent Gaussian chaotic–Lorentzian fields. Thus the superposition of chaotic fields has a nice interpretation in terms of the dynamics of the cavity evolution; a non-Markov evolution characterized by age-dependent emission and absorption rates produces radiation whose second-order characteristics coincide with those of a thermal field obtained by the superposition of two independent thermal fields, each with a Lorentzian spectrum.

The identification assumes special importance for two reasons. First, the derivation of the statistical characteristics of the radiation resulting from superposition is a fairly difficult mathematical problem, even in the case of chaotic fields. The method of analysis presented in this chapter can be adapted to consider even more intricate situations. Second, there is an interpretation of the origin of the non-Markov nature of the chaotic field in question. For instance, the condition 8.3.34 or 8.3.36 characterizing the Bose–Einstein fluctuation can be dispensed with; then we have a more general field which is non-Gaussian. The analysis will lead to the determination of the second-order properties of the field.

8.4 Anti-bunching and sub-Poissonian characteristics

We finally show that the non-Markov evolution modelled and dealt with so far can also lead to anti-bunched and sub-Poissonian photon statistics. The part of amplification process of the field corresponding to spontaneous emission has been modelled as a Poisson process which in turn implies Markov evolution. If this is also modelled as a non-Markov process, then it can usually be expected to result in the reduction of bunching; at any rate we have seen some examples in Chapters 4 and 7. We shall accordingly model the immigration process (spontaneous emission) as in section 7.5 as a process of delayed Poisson events, the delay itself being an exponentially distributed random variable. In other words, spontaneous emissions normally occur at an (expected) rate v per unit of (coarse-grained) time; however, after each emission, there is a dead time distributed exponentially with parameter $\beta > 0$, so after the expiry of the dead time spontaneous emissions can occur. The time to the (first) emission is distributed exponentially with parameter v, the dead time again coming into operation immediately after emission. The emissions are as indicated in Fig. 7.5.2, page 128. The process of absorption by cavity and stimulated emission is modelled exactly as in section 8.1. Of course the analysis can be carried through for the asymmetric model of section 8.3; however, we wish to consider a simple model so that the computations are not unduly complex. Thus the analysis and results of section 8.1, insofar as they relate to the functions $g_i(z_1, z_2, z_3, t)$ $(i = 1, 2)$, are valid in the present model; however, equations 8.1.3 through 8.1.5 are no longer valid and have to be replaced by new ones.

As in Chapter 7, it is convenient to introduce the two-valued process $\{Z(t)\}$ taking values 0 and 1 corresponding to the dead and live phases of the process of spontaneous emissions. We introduce the generating function $G_i(z_1, z_2, z_3, z, t)$ $(i = 0, 1)$ by*

---

* The functions G_i are different from the ones introduced in 8.3; no confusion is possible since we shall not deal with the functions G, G_1, G_2 introduced in section 8.3.

$$G_i(z_1, z_2, z_3, z, t)$$

8.4.1
$$= E\left[z_1^{X_1(t)} z_2^{X_2(t)} z_3^{X_3(t)} z^{W(t)} \,\middle|\, W(0) = 0,\, v \neq 0,\, Z(0) = i\right].$$

By analysis of the backward differential relation, we obtain

8.4.2
$$\frac{\partial G_0}{\partial t}(z_1, z_2, z_3, z, t) = \beta G_0(z_1, z_2, z_3, z, t)$$
$$+ \beta G_1(z_1, z_2, z_3, z, t)$$

8.4.3
$$\frac{\partial G_1}{\partial t}(z_1, z_2, z_3, z, t) = -v G_1(z_1, z_2, z_3, z, t)$$
$$+ v G_0(z_1, z_2, z_3, z, t) \times g_1(z_1, z_2, z_3, z, t),$$

with the initial conditions

8.4.4
$$G_i(z_1, z_2, z_3, z, 0) = 1 \quad (i = 0, 1).$$

To obtain the moments we introduce the following notation:

8.4.5
$$A_i(t) = \frac{\partial G_i}{\partial z}(a_1, z_2, z_3, z, t)\Big|_{z = z_1 = z_2 = z_3 = 1} \quad i = 0, 1$$

8.4.6
$$B_i(t) = \frac{\partial^2 G_i}{\partial z^2}(z_1, z_2, z_3, z, 1)\Big|_{z = z_1 = z_2 = z_3 = 1} \quad i = 0, 1.$$

In addition to the moments and cross-correlations $a_k^i(t)$, $a^i(t)$, $b_k^{ij}(t)$, $b^{ij}(t)$ defined by 8.1.10 and 8.1.11, we introduce the following further notation:

8.4.7
$$a_k(t) = \sum_{i=1}^{3} a_k^i(t), \quad b_k(t) = \sum_{i,j=1}^{3} b_k^{ij}(t).$$

Differentiating 8.4.2 and 8.4.3 w.r.t. z and evaluating at $z_i = z = 1$, we obtain

8.4.8
$$\frac{dA_0(t)}{dt} = -\beta A_0(t) + \beta A_1(t)$$

8.4.9
$$\frac{dA_1(t)}{dt} = -v A_1(t) + v(A_0(t) + a_1(t))$$

8.4.10
$$\frac{dB_0(t)}{dt} = -\beta B_0(t) + \beta B_1(t)$$

8.4.11
$$\frac{dB_1(t)}{dt} = -vB_1(t) + v(B_0(t) + b_1(t)) + 2vA_0(t)a_1(t),$$

with the initial conditions

8.4.12
$$A_i(0) = B_i(0) = 0 \quad i = 0, 1.$$

Using Laplace transform technique, we obtain

8.4.13
$$A_0^*(s) = \frac{\beta v a_1^*(s)}{s(s+\beta+v)}, \quad A_1^*(s) = \frac{(\beta+s)v a_1^*(s)}{s(s+\beta+v)}$$

8.4.14
$$B_0^*(s) = \frac{\beta v(b_1^*(s) + L(s))}{s(s+\beta+v)}, \quad B_1^*(s) = \frac{(\beta+s)v(b_1^*(s) + L(s))}{s(s+\beta+v)}$$

where $L(s)$ is given by

8.4.15
$$L(s) = 2 \int_0^\infty A_0(t)a_1(t)\,e^{-st}\,dt.$$

We need only the value $L(0)$ and it can be obtained either directly or following the method suggested in section 7.5; we obtain after some computations

$$L(0) = \frac{\beta v}{2(\lambda+\mu+\eta)[(\beta+\lambda+v+\mu+\eta)^2 - \lambda\alpha]}$$
$$\{2\lambda+\mu+\eta)^2[3(\lambda+\mu+\eta)^2 + \lambda\alpha + 2(\beta+v)(\lambda+\mu+\eta)]/$$

8.4.16
$$[(\lambda+\mu+\eta)^2 - \lambda\alpha]^2 - 1\}$$

We are now comfortably placed to obtain the first two moments of the equilibrium cavity photon population. As usual we neglect the population in phase 3; thus the first moment is given by

8.4.17
$$\lim_{t \to \infty} A_0(t) = \lim_{t \to \infty} A_1(t) = \beta v a_1^*(0)/(\beta+v),$$

where $a_1^*(0)$ can be evaluated using 8.4.7 and 8.1.18 through 8.1.20:

8.4.18
$$a_1^*(0) = (2\lambda+\mu+\eta)/[(\lambda+\mu+\eta)^2 - \lambda\alpha].$$

The second factorial moment is evaluated in a similar way:

8.4.19
$$B_0(\infty) = B_1(\infty) = \beta v[b_1^*(0) + L(0)]/(\beta+v)$$

8.4.20
$$b_1^*(0) = \lambda\alpha(3\lambda+2\mu+2\eta+\alpha)/2([\lambda+\mu+\eta]^2 - \lambda\alpha]^2.$$

As usual introducing the bunching factor $\mathscr{B}$ by

8.4.21 $\qquad \mathscr{B} = B_i(\infty)/[A_i(\infty)]^2,$

we find

$$\mathscr{B} = \frac{(\beta + v)}{2\beta v(2\lambda + \mu + \eta)^2}\left\{\lambda\alpha(3\lambda + 2\mu + 2\eta - \alpha) + \right.$$

$$+ \beta v\left[(2\lambda + \mu + \eta)^2\left(3\lambda + 3\mu + 3\eta + 2\beta + 2v + \frac{\lambda\alpha}{\lambda + \mu + \eta}\right) - \right.$$

8.4.22 $\qquad \left.\left. - \frac{[(\lambda + \mu + \eta)^2 - \lambda\alpha]^2}{\lambda + \mu + \eta}\right]\middle/ [(\beta + v + \lambda + \mu + \eta)^2 - \lambda\alpha]\right\}.$

If we choose $\mu + \eta = 2\lambda$, $\alpha = \lambda/2$, $v = \lambda$, $\beta = \lambda/2$, we find $\mathscr{B} = 0.66$, showing that the photon statistics is anti-bunched and sub-Poissonian in its character. If we wish to obtain the product densities, we need follow the method outlined in section 7.6. Since the method of analysis is identical, we do not discuss it any further.

To sum up: the non-Markov modelling used in this chapter is versatile enough to bring out all the characteristics of the cavity population from thermal behaviour on the one hand to anti-bunched photon statistics on the other.

Additional notes

There is a huge literature on thermal light and its detection. The collection of papers edited by Mandel and Wolf (1970) contains a good number of papers dealing with thermal light. Besides, Saleh (1978) has provided a detailed account of thermal light and the photoelectron statistics arising therefrom. For additional references, the bibliography provided by Saleh may be consulted. The analysis and results relating to non-Markov evolution are based on the recent contributions by the author (1986d, e). Photon-counting statistics of thermal light consisting of two independent spectral lines had been studied by Sukavanam (1973, 1981) and Tornau and Echtermeyer (1973); these analyses are essentially based on the representation of the photon-counting process as a Cox process, as explained in section 3.4.

9. A brief summary and outlook

As we have remarked earlier, direct stochastic modelling of cavity radiation and detection is interesting from many points of view; in this monograph we have attempted to characterize the features of the assembly of photons by the population point process approach. After a short introduction and presentation of a summary of the main results of stochastic processes, we have outlined the method of dealing with cavity population with special reference to problems of detection. Then we have highlighted the physics of cavity evolution and detection of the resulting radiation, a process which has led us to stochastic evolution. Since the final evolution equation depends very much on the time-scale adopted and the consequent coarse-graining, we have taken as the convenient starting-point features relating to oscillation below threshold which strongly favour a branching type evolution. A strictly Markov type of evolution combined with the branching behaviour makes the resulting radiation thermal with a Lorentzian spectrum. The analysis and results discussed in Chapters 7 and 8 mainly relate to models in which the Markov character has been dispensed with. To make the point clear we start by analyzing the features in more detail.

In the Markov model, we have visualized the field as an assembly of photons which we call cavity population. The population of cavity atoms and its interaction with the field is taken into account in the following manner: each photon, independent of other photons, has a constant probability λ per unit time (in the coarse-grained picture) of producing another photon and a constant probability μ per unit time of being absorbed by the cavity atoms. Thus the population of atoms is accorded secondary importance. These two assumptions account for the phenomena of absorption and stimulated emission; however, from a probabilistic point of view, the population either becomes extinct ($\mu > \lambda$) or explodes ($\mu < \lambda$). Thus to obtain a non-trivial steady state of equilibrium, it is necessary to support the process by an immigration. To conform to Markov type of behaviour, it is necessary to postulate that the (expected) rate of immigration is a constant equal to v. If this is done, then the steady-state population size has a negative binomial distribution when $\lambda < \mu$. In fact, this is the model of population of cavity photons, first proposed by Shimoda, Takahasi and Townes. In order that the distribution may reduce to Bose–Einstein, it is necessary that $\lambda = v$ which is also in conformity with the second quantized picture in which the probability of emission is proportional to $n + 1$ when the population size is n. In such a case, Shimoda *et al.* interpret the ratio in terms of the effective temperature T_e defined by

$$9.1 \qquad \frac{\lambda}{\mu} = \exp -(\hbar\varepsilon/kT_e)$$

where $\hbar\varepsilon$ is the energy gap between the upper and lower levels and k is the Boltzmann constant.

Next we examine the effects of the detector. This is done by assuming that there is an emigration process, the (expected) rate of emigration per photon being a constant equal to η. The emigration process does not disturb the Markov nature of the enlarged system of cavity, field and detector. The constancy requirement of the rate is due to the fast nature of the detector;* in other words, the detector atoms absorb the photons much more readily than for instance the cavity atoms.

In any case the introduction of the detector disturbs the system; but soon the enlarged system of the cavity, field and detector reaches equilibrium, as demonstrated in Chapter 6; equation 9.1 being valid in this case with the replacement $\mu \mapsto \mu + \eta$. The spectrum of the resulting radiation as recorded by the detector corresponds to the autocorrelation $R(t)$, where

$$9.2 \qquad R(t) = \frac{\lambda}{\mu + \eta - \lambda} \exp -\{(\mu + \eta - \lambda)/2\}|t|,$$

as has been demonstrated in section 6.3. Viewed from the broad framework of quantum detection, this corresponds to the field being Gaussian Markov. Since the Gaussian field corresponds to chaotic radiation, the identification of the autocorrelation by equation 9.2 singles out the model corresponding to the Markov field. This has led us naturally into the quest for non-Markov evolution.

---

* Even in the case of radiation due to non-Markov evolution, the constancy of the rate of detection should not be dispensed with, unless, of course, we deliberately resort to some kind of slow counting, in which case the detector may not be faithful and the spectrum as shown by the detector may have to be corrected before being identified as the one corresponding to the field (see for example Srinivasan and Vasudevan (1985)).

The models and results presented in Chapters 7 and 8 correspond to non-Markov evolution; the division corresponding to the two different types of evolution, although convenient from the point of view of analysis of the models, is rather unphysical. We have seen in section 7.5 that the particular type of non-Markov evolution can be thought of either as a Bellman–Harris process or as a Kendall process of population growth. This is interesting because the interpretation of age-dependent birth- and death-rates in terms of real photons has no physical basis, and the fact that a particular model can be thought of as evolving according to either type of age-dependence only goes to emphasize that it is only a device in the mathematical analysis of the model and that the concept of age-dependence should be de-emphasized in favour of non-Markov evolution which is more easily interpretable in a physical sense. The results that are derived in general and the correlation functions of the detection process in particular bring to the fore the viability of the non-Markov models, as is to be expected. The most interesting features are:

(i)   chaotic radiation with a Lorentzian spectrum can be extracted as a special case of the non-Markov evolution of cavity population of photons;

(ii) chaotic radiation corresponding to the superposition of independent Gaussian Lorentzian fields is accommodated (to at least second order) in the non-Markov formulation, particularly the one using multiphase approach;

(iii) radiation with anti-bunched statistics is also accommodated within the framework of non-Markov formulation, particularly when the process of spontaneous emission is appropriately modelled as a non-Markov point process.

The last two features throw additional light on the problem of evolution and detection in general. For instance, the superposition of fields admits a direct dynamical interpretation; the population evolution has to exhibit necessarily non-Markov character, maybe on a coarse-grained time-scale. Likewise, anti-bunching, although a consequence of quantum theory, is characterized by some kind of break in spontaneous emission, with or without a corresponding property in the case of simulated emission. Of course the first feature bringing the thermal Lorentzian profile within the ambit of non-Markov evolution is rather puzzling. In the first instance we might argue that the equivalence is only up to second order; however, this also can be rather seriously challenged, particularly in view of the results in section 8.2c relating to the third moment. Viewed this way, we can end in a satisfactory vein by emphasizing the rather viable nature of non-Markov evolution in general.

As we have discussed in Chapter 5, the emission process of photoelectrons can be thought of as being either

(i) an emigration process arising out of a cavity population process, or

(ii) a Cox process with intensity identified as $\alpha E^+(t)E^-(t)$.

The identification (i) follows from the evolution equation satisfied by the diagonal elements of the density matrix (over coarse-grained time), while (ii) is a natural consequence of the definition of coherence functions. The Markov model in picture (i) leads to the identification of the Cox process with the stationary point process of emigration. In section 6.2, we have seen that even if the stationary equilibrium condition is not imposed, in some cases it is possible to identify the emigration process as a Cox process. The results of Chapter 8 go a little further in this direction in that the non-Markov evolution of the population also leads to an emigration process that corresponds to a Cox process corresponding to a Gauss–Markov field. Detailed investigation may lead to a deeper result in the theory of stochastic processes. To the best of my knowledge, even the identification discussed in section 6.2 (see 6.2.9 through 6.2.11) has not been noted by probabilists, although there are other instances in the theory of stochastic processes where a non-Markov process can be approximated as a Markov process (see Hasminskii (1980)).

In this monograph we have not paid any attention to other features like phase which can play an important role in a phenomenon admittedly quantum in nature. To take into account these features, it may be necessary to study coarse-graining with special reference to the elements of the density matrix that are not necessarily diagonal. Maybe a kind of non-Markov quantum evolution equations need to be developed, the elements of which are already implicit in the semi-quantum or fully quantum version of the operator equations developed by Haken (1970), Lax (1967), and Louisell (1973). We may have to deal with the stochastic differentials for the second quantized operators and include the effects of coarse-graining by some kind of support terms as envisaged by Haken (1970). Such a procedure is by no means easy and the differential equations may be formidably difficult in view of the unquantized stochastic source terms. However, in the limited context of optical up-conversion (see for example Mielniczuk (1979) and Srinivasan (1984)) exact results relating to the values of the Green's functions can be extracted. This is rather encouraging, and further research into modelling by resort to a convenient set of equations for the operators themselves may be fruitful.

References

Athreya, KB and Ney, PE (1972)   *Branching Processes* (Berlin: Springer-Verlag)

Bartlett, MS (1963)   Spectral analysis of point processes. *J. Roy. Statist. Soc.* **B 25**, 261–96

Bartlett, MS (1975)   *Stochastic Processes*, 3rd edn (Cambridge: University Press)

Bellman, RE and Harris, TE (1948)   On the theory of age dependent stochastic processes. *Proc. Nat. Acad. Sci. USA* **34**, 601–4

Bhabha, HJ (1950)   On the stochastic theory of continuous parametric systems and its application to electron–photon cascades. *Proc. Roy. Soc.* **A 202**, 300–32

Bloembergen, N (1956)   Proposal for a new type solid state maser. *Phys. Rev.* **104**, 324–7

Bloembergen, N (1965)   *Non-linear Optics* (New York: Benjamin)

Bonifacio, R, Schwendimann, P and Haake, F (1971)   Quantum theory of super-radiance I and II. *Phys. Rev.* **A 4**, 312–13, 854–64

Bosov, NG and Prokhorov, AM (1954)   Application of molecular beams to the radio spectroscopic study of the rotation spectra of molecules. *J. Exptl. Theor. Phys. (USSR)* **27**, 431–8 (in Russian)

Campbell, N)1909)   The study of discontinuous phenomena. *Proc. Camb. Phil. Soc.* **15**, 117–36

Chung, KL (1967)   *Markov Chains with Stationary Transition Probabilities*, 2nd edn (Berlin: Springer-Verlag)

Cox DR (1962)   *Renewal Theory* (London: Methuen)

Cox, DR and Lewis, PAW (1966)   *The Statistical Analysis of Series of Events* (London: Methuen)

Cox DR and Lewis PAW (1972)   Multivariate point processes, in *Proceedings of the 6th Berkeley Symposium on Probability and Mathematical Statistics*, Ed. J. Neyman and EL Scott (Berkeley: University of California)

Daley DJ and Vere-Jones D (1972)   A summary of the theory of point processes, pp. 299–383 in Lewis (1972)

Dyson FJ and McVoy KW (1957)   Longitudinal polarization of bremsstrahlung from polarized electrons in Born approximation. *Phys. Rev.* **106**, 828–9

Einstein A (1917)   The quantum theory of radiation. *Physikal. Zeits.* **18**, 121–8

Feller W (1941)   On the integral equation of renewal theory. *Ann. Math. Statist.* **12**, 243–67

Feller W (1949)   Fluctuation theory of recurrent events. *Trans. Amer. Math. Soc.* **67**, 98–119

Feller, W (1968)   *An Introduction to Probability Theory and its Applications*, vol. 1 (New York: John Wiley)

Feller, W (1971)   *An Introduction to Probability Theory and its Applications*, vol. 2 (New York: John Wiley)

Feynman, RP and Vernon, FL (1963)   The theory of a general quantum system interacting with a linear dissipative system. *Ann. Phys.* **24**, 118–73

Glauber, RJ (1963)   Coherent and incoherent states of the radiation field. *Phys. Rev.* **131**, 2766–88

Glauber, RJ (1969)   Coherence and quantum detection, in *Proceedings of the International School of Physics "Enrico Fermi", Course XII Quantum Optics*, ed. RJ Glauber, pp. 15–56 (New York: Academic Press)

Glauber, RJ (1970)   Quantum theory of coherence, in *Quantum Optics*, ed. SM Kay and A Maitland, 53–125 (New York: Academic Press)

Gnedenko, BV (1962)   *The Theory of Probability*, 2nd edn (New York: Chelsea)

Gordon JP, Zeiger, HZ and Townes, CH (1954)   Molecular oscillator and new hyperfine structure in the microwave spectrum of NH_3. *Phys. Rev.* **95**, 282–4

Haken, H (1970)   *Laser Theory, Handbuch der Physik* **25**-2c (Berlin: Springer-Verlag)

Hanbury-Brown, R and Twiss, RQ (1954)   A new type of interferometer for use in radio astronomy. *Phil. Mag.* **45**, 663–82

Hanbury-Brown, R and Twiss, RQ (1957)   Interferometry of the intensity fluctuations in light: 1. *Proc. Roy. Soc. (Lond.)* **A242**, 300–24

Harding, EF and Kendall, DG (1973)   *Stochastic Analysis* (New York: Wiley)

Harris, TE (1962)   *The Theory of Branching Processes* (Berlin: Springer-Verlag)

Hasminskii, RZ (1980)   *Stochastic Stability of Differential Equations*. Transl. by D Louvish (Alphen aan den Rijn, The Netherlands: Sijthoff & Noordhoff)

Huang, XY, Narducci, LM and Yuan, JM (1981)   Population equations for quantum systems in contact with dissipation mechanisms. *Phys. Rev.* **A 23**, 3084–93

Jaggers, P (1975)   *Branching Processes with Biological Applications* (New York: Wiley)

Janossy, L (1950a)   On the absorption of a nucleon cascade. *Proc. Roy. Irish Acad. Sci.* **A 53**, 181–8

Janossy, L (1950b)   Note on the fluctuation problem of cascades. *Proc. Phys. Soc.* **A 63**, 241–9

Kallenberg, O (1975)   *Random Measures* (Berlin: Akademie-Verlag)

Karlin, S (1966)   *A First Course in Stochastic Processes* (New York: Academic Press)

Kendall, DG (1949)   Stochastic processes and population growth. *J. Roy. Statist. Soc.* **B 11**, 230–64

Kendall, DG (1966)   Branching processes since 1873. *J. Lond. Math. Soc.* **41**, 385–406

Kendall, DG and Harding, EF (Eds) (1974)   *Stochastic Geometry* (New York: Wiley)

Khintchine, AY (1955)   *Mathematical Methods in the Theory of Queueing* (Russian). Transl. by DM Andrews and MH Quenouille (London: Griffin)

Kimble, HJ, Dagenais, M and Mandel, L (1977)   Photon antibunching in resonant fluorescence. *Phys. Rev. Lett.* **39**, 691–5

Kimble HJ, Dagenais, M and Mandel, L (1978)   Multiatom and transit-time effects on photon-correlation measurements in resonant fluorescence. *Phys. Rev.* **A 18**, 201–7

Klauder, JR and Sudarshan, ECG (1968)   *Fundamentals of Quantum Optics* (New York: Benjamin)

Kolmogorov, AN and Dmitriev, NA (1947)   Branching stochastic processes. *Doklady* **56**, 5–8

Kuznetsov, PI and Stratonovic, RL (1956)   A note on mathematical theory of correlated random points. Transl. by JA McFadden in *Selected Translations in Mathematical Statistics and Probability* **7** (1968) 1–16 (Providence: American Mathematical Society)

Lamb, WE (1964)   Theory of an optical maser. *Phys. Rev.* **B 4**, 1429–50

Lax, M (1967)   Quantum noise X: Density matrix treatment of field and population difference fluctuations. *Phys. Rev.* **157**, 213–31

Lax, M and Louisell, WH (1967)   Quantum noise IX: Quantum Fokker–Planck equation for laser noise. *IEEEJ Quantum Electronics* **QE-3**, 47–58

Lewis, PAW (Ed) (1972)   *Stochastic Point Processes: Statistical Analysis, Theory and Applications* (New York: Wiley)

Loudon, R (1980)   Non-classical effects in the statistical properties of light. *Rep. Prog. Phys.* **43**, 913–49

Louisell, WH (1964)   *Radiation and Noise in Quantum Electronics* (New York: McGraw-Hill)

Louisell, WH (1973)   *Quantum Statistical Properties of Radiation* (New York: John Wiley)

McFadden, JA (1962)   On the lengths of intervals in a stationary point process. *J. R. Statist. Soc.* **B 24**, 364–82

Mandel, L (1958)   Fluctuations of photon beams and their correlations. *Proc. Phys. Soc.* **72**, 1037–48

Mandel, L (1959)   Fluctuations of the photon beams: The distribution of photoelectrons. *Proc. Phys. Soc.* **74**, 233–43

Mandel, L (1979)   Sub-Poissonian photon statistics in resonance fluorescence. *Opt. Lett.* **4**, 205–7

Mandel, L, Sudarshan, ECG and Wolf, E (1964)   Theory of photoelectric detection of light fluctuations. *Proc. Phys. Soc.* **A 84**, 435–44

Mandel, L and Wolf, E (Eds) (1970)   *Selected Papers on Coherence and Fluctuations of Light*, Vol. 1 and 2 (New York: Dover)

Mathes, K (1978)   *Infinitely Divisible Point Processes* (New York: Wiley)

Mielniczuk, WJ (1979)   An exactly soluble model of optical up-conversion with stochastic pumping. *Optica Acta* **26**, 1115–19

Mode, CJ (1971)   *Multitype Branching Processes: Theory and Applications* (New York: Elsevier)

Moullins, EB (1938)   *Spontaneous Fluctuations of Voltage* (due to Brownian motions of electricity, shot effect and kindred phenomena) (Oxford: University Press)

Moyal, JE (1962)   General theory of stochastic population processes. *Acta Math.* **108**, 1–31

Murthy, VK (1974)   *General Point Processes—Applications to Structural Fatigue, Bioscience and Medical Research* (Reading, Pa.: Addison-Wesley)

Palm, C (1943)   Intensitätschwankungen in Fernsprechvorkehr. *Ericsson Technics* **44**, 1–189

Parzen, E (1960)   *Modern Probability Theory and its Applications* (New York: John Wiley)

Parzen, E (1962)   *Stochastic Processes* (New York: Holden-Day)

Paul, H (1982)   Photon antibunching. *Rev. Mod. Phys.* **54**, 1061–1102

Ramakrishnan, A (1950)   Stochastic processes relating to particles distributed in a continuous infinity of states. *Proc. Camb. Phil. Soc.* **46**, 595–602

Ramakrishnan, A (1953)   Stochastic processes associated with random divisions of a line. *Proc. Camb. Phil. Soc.* **49**, 473–85

Ramakrishnan, A (1958)   Probability and stochastic processes, in *Handbuch der Physik*, vol. 4 (Berlin: Springer-Verlag)

Ramakrishnan, A and Srinivasan, SK (1956)   A new approach to cascade theory. *Proc. Ind. Acad. Sci.* **A 44**, 263–73

Rice, SO (1945)   Mathematical analysis of random noise. *Bell Sys. Tech. J.* **25**, 46–156

Riesz, F and Nagy, BSz (1955)   *Functional Analysis* (New York: Frederick Unger)

Risken, H (1984)   *The Fokker–Planck Equation: Methods of Solution and Applications* (Berlin: Springer-Verlag)

Rowland, EN (1936)   The theory of mean square variation of a function formed by adding known functions with random phases and application to the theories of the shot effect and of light. *Proc. Camb. Phil. Soc.* **32**, 580–97

Rowland, EN (1937)   The theory of shot effect: II. *Proc. Camb. Phil. Soc.* **33**, 344–58

Rowland, EN (1938)   Note on fluctuations and the shot effect. *Proc. Camb. Phil. Soc.* **34**, 329–44

Saleh, BEA (1978)   *Photo-electron Statistics* (Berlin: Springer-Verlag)

Sampath, G and Srinivasan, SK (1977)   *Stochastic Models for Spike Trains of Single Neurons* (Berlin: Springer-Verlag)

Schell, A and Barakat, R (1973)   Approach to equilibrium of single mode radiation in a cavity. *J. Phys.* **A 6**, 826–36

Schwinger, J (1960)   Field theoretic methods in non-field theoretic context, in *Brandeis University Summer Institute in Theoretical Physics* (Brandeis: University Press)

Scully, MO and Lamb, WE (1967)   Quantum theory of an optical maser. *Phys. Rev.* **159**, 208–26

Scully, MO and Lamb, WE (1968) Quantum theory of an optical maser, II: Spectral profile. *Phys. Rev.* **166**, 246–9

Scully, MO and Lamb, WE (1969) Quantum theory of an optical maser, III: Theory of photoelectron counting statistics. *Phys. Rev.* **179**, 368–74

Shepherd, TJ (1981) A model for photodetection of single-mode cavity radiation. *Optica Acta* **28**, 567–83

Shimoda, K, Takahasi, H and Townes, CH (1957) Fluctuations in amplification of quanta with application to maser amplifiers. *J. Phys. Soc. Japan* **12**, 686–700

Short, R and Mandel, L (1983) Observation of sub-Poissonian photon statistics. *Phys. Rev. Lett.* **51**, 384–7

Smith, WL (1954) Asymptotic renewal theorems. *Proc. Roy. Soc. Edin.* **A 64**, 9–48

Smith, WL (1958) Renewal theory and its ramifications. *Proc. Roy. Statist. Soc.* **B 20**, 243–302

Snyder, DL (1975) *Random Point Processes* (New York: Wiley Interscience)

Srinivasan, SK (1965) A novel approach to the theory of shot noise. *Nuovo Cimento* **38**, 979–92

Srinivasan, SK (1969) *Stochastic Theory and Cascade Processes* (New York: Elsevier)

Srinivasan, SK (1974a) *Stochastic Point Processes and their Applications* (London: Griffin)

Srinivasan, SK (1974b) Photocount statistics of Gaussian light of time limited spectral profile. *Phys. Lett.* **A 47 A**, 151–2

Srinivasan, SK (1974c) Dead time effects in photon counting statistics. *Phys. Lett.* **58 A**, 277–8

Srinivasan, SK (1978) Dead time effects in photon counting statistics. *J. Phys.* **A 11**, 2333–40

Srinivasan, SK (1984) A canonical model of optical up-conversion with stochastic pumping exact results. *Optica Acta* **31**, 785–93

Srinivasan, SK (1986a) Analysis of characteristics of light generated by a space charge limited electron stream. *Optica Acta* **33**, 207–11

Srinivasan, SK (1986b) Generation of antibunched light by an inhibited Poisson stream. *Optica Acta* **33**, 835–42

Srinivasan, SK (1986c) A model of cavity radiation with antibunched and sub-Poissonian characteristics. *J. Phys.* **A 19**, L595–8

Srinivasan, SK (1986d) A non-Markov model of cavity radiation with thermal characteristics. *J. Phys.* **A 19**, L513–16

Srinivasan, SK (1986e) An age dependent model of cavity radiation and detection. *Pramana—J. of Physics* **27**, 19–31

Srinivasan, SK (1987) A non-Markov model of cavity radiation and its detection. *J. Mod. Opt.* **34**, 291–306

Srinivasan, SK and Subramanian, R (1980) *Probabilistic Analysis of Redundant Systems* (Berlin: Springer-Verlag)

Srinivasan, SK and Sukavanam, S (1971) Photo count statistics of gaussian light of arbitrary spectral profile. *Phys. Lett.* **35 A**, 81–2

Srinivasan, SK and Sukavanam, S (1972) Photo count statistics of gaussian light with arbitrary spectral profile. *J. Phys.* **A 5**, 682–94

Srinivasan, SK, Sukavanam, S and Sudarshan, ECG (1973) Many-time photocount distributions. *J. Phys.* **A 6**, 1910–18

Srinivasan, SK and Vasudevan, R (1985) A non-Markovian model of photo-detection of cavity radiation. *Optica Acta* **32**, 749–66

Srinivasan, SK and Vasudevan, R (1986a) A non-Markov model of cavity radiation and its detection. *Optica Acta* **33**, 191–205

Srinivasan, SK and Vasudevan, R (1986b) Approach to equilibrium of single mode cavity radiation. *J. Math. Anal. Applicns* **119**, 249–58

Sudarshan, ECG (1963) Equivalence of semi classical and quantum-mechanical descriptions of statistical light beams. *Phys. Rev. Lett.* **10**, 277–9

Sukavanam, S (1973) Some problems in photon-counting statistics. Ph.D. Thesis, Indian Institute of Technology, Madras

Sukavanam, S (1981) Photo-electron counting statistics of partially polarized light beams. *J. Math. Phys. Sci.* **15**, 47–58

Teich, MC and Saleh, BEA (1985) Observation of sub-Poisson Franck–Hertz light 253.7 nm. *J. Opt. Soc. America* **2**, 275–82

Teich, MC and Saleh, BEA (1987) Photon bunching and anti-bunching. *Prog. in Optics.* Ed. E Wolf **25** (in press)

Teich, MC, Saleh, BEA and Stoler, D (1983) Antibunching in the Frank–Hertz experiment. *Opt. Comm.* **46**, 244–8

Titchmarsh, EC (1937) *Introduction to the Theory of Fourier Integrals.* (Oxford: University Press)

Tornau, N and Echtmayer, B (1973) Photon counting statistics of thermal light consisting of two spectral lines. *Ann. der Physik* **29**, 289–301

Vere-Jones, D (1970) Stochastic models for earthquake occurrence. *J. Roy. Statist. Soc.* **B 32**, 1–62

Von Mises, R (1936) La loi de probabilité pour les fonctions statistiques. *Ann. Inst. H. Poincaré* **6**, 185–209

Walker, JG and Jakeman, E (1985) Photon anti-bunching by use of a photo electron-event-triggered optical shutter. *Optica Acta* **32**, 1303–8

Watson, HW and Galton, F (1874) On the probability of extinction of families. *J. Anthropol. Inst. Great Britain and Ireland* **4**, 138–44

Index

Amplification, 70
Antibunching, 137, 142, 164, 166
Athreya, KB, 69

Backward equation, 14
Barakat, R, 110
Bartlett, MS, 22, 35, 50
Bellman, RE, 69
Bellman–Harris process, 59–61, 170
   modified, 112–42
Bhabha, HJ, 49
Black-body radiation, 78, 79
Bloembergen, N, 2
Bonifacio, R, 142
Bosov, NG, 2
Bunching factor, 101, 156

Campbell, N, 49
Cavity evolution, 89
—field, density matrix, 71
— —, evolution equation, 76
— —, quantum evolution, 71, 77
—gain, 75
—loss, 75, 76
—radiation, 70ff.
Chapman–Kolmogorov equation, 13
Chung, KL, 13
Coarse graining, 76, 77
Coherent model, 123
—states, 84
Coincidence functions, 85
Cox, DR, 36, 50, 51, 119
Cox processes, 42, 43, 50, 167, 171

Dagenais, M, 142
Daley, DJ, 51
Dead time, 47
  corrections, 71
Density matrix, 71
Detection process, 79–83, 93–105
Dimitriev, M, 64
Dissipation, 75
Dyson, FJ, 51

Echtmayer, B, 167
Einstein, A, 87
Electron–photon cascade, 32
Energy straggling, 20

Feller, W, 13, 20, 50
Feynman, RP, 2
Forward equation, 14

Galton, F, 69
Glauber, RJ, 2, 84
Gnedenko, BV, 4
Gordon, JP, 2, 87

Haake, F, 142
Haken, H, 2, 88, 171
Hanbury-Brown, R, 2, 88
Harding, EF, 51
Harris, TE, 51, 69, 114
Hasminskii, RZ, 171
Huang, XY, 142

Inclusive probabilities, 85

Jaggers, P, 69
Jakeman, E, 142
Jonossy, L, 50

Kallenberg, O, 51
Karlin, S, 22
Karuhnen–Loeve expansion, 8
Kendall, DG, 49, 51, 69
Kendall process, 61–4, 170
   modified, 64–8, 144–67
Khintchine, AY, 49
Kimble, HJ, 142
Klauder, JR, 88
Kolmogorov, AN, 69
Korolyuk's theorem, 37
Kuznetsov, PI, 50

Lamb, WE, 2, 79, 80, 88
Laser, 70, 78
Lax, M, 2, 88, 171
Lewis, PAW, 51
Loudon, R, 142
Louisell, WH, 2, 88, 171
Mandell, L, 2, 82, 88, 142, 167
Markov field, 169
—process, 11
Maser, 2
Mathes, K, 51
McFadden, JA, 50

McVoy, KW, 51
Mercer's theorem, 8
Mielniczuk, JW, 171
Mode, CJ, 69
Moment measure, 26
Moullins, EB, 48
Moyal, JE, 50
Multiplicative chain, 20
Murthy, VK, 51

Nagy, BSZ, 8
Narducci, LM, 142
Negative binomial distribution, 56, 90, 91
Ney, PE, 69

Parzen, E, 4, 22
Paul, H, 143
Phase approach, 63–8, 144–52
Photoemission, 86, 99, 127, 138, 152
   dead-time corrected, 105
Point process, 23–51
   birth, 29, 38
   cavity population, 79
   death, 31, 38
   detection, 79–83, 93–105, 116–18
   emigration, 56
   immigration, 52
   marked, 31
   multidimensional, 50
   multivariate, 31
   Poisson, 36
   population, 23
   regular, 37
   renewal, 36, 41, 42
   stationary, 35, 36
Poisson process, 6, 14, 36
Process:
   age-dependent, 59, 61
   birth-and-death, 17, 28
   Bellman–Harris, 59–61, 112–42, 170
   branching, 20
   cavity evolution, 76, 77
   complex-Gaussian, 7
   compound, 26
   counting, 25
   covariant stationary, 6
   Cox, 42, 43, 50, 167
   detection, 79–83, 93–105, 116–18
   doubly stochastic Poisson, 42, 43, 50
   emigration, 52, 56
   Feller–Kolmogorov, 20
   Gaussian, 7
   Kendall, 61

Process: (continued)
   Markov, 11
   Poisson, 6, 14, 36
   population growth, 52–69
   purely discontinuous, 20
   renewal, 36
   self-exciting, 43
   self-inhibitive, 43
   stationary, 5
   stochastic, 4
   strictly stationary, 5
   two-state, 15
Product density, 27, 28
   conditional, 30
   multivariate, 31
Prokhorov, AM, 2

Quantum detection, 83–7
— evolution, 76

Radiation:
   black-body, 78–9
   cavity, 70
   chaotic, 79
Ramakrishnan, A, 22, 49, 50, 51
Random dead-time model, 128
Random telegraph signal, 16
Recurrence time:
   backward, 36
   forward, 36
Rice, SO, 49
Riesz, F, 8
Risken, H, 88
Rowland, EN, 49

Saleh, BEA, 88, 142, 143, 167
Sampath, G, 51
Schell, A, 110
Schwendimann, 142
Schwinger, J, 2
Scully, MO, 2, 79, 80, 88
Shepherd, TJ, 111
Shepherd model, 111, 119
Shimoda, K, 2, 110
Short, R, 142
Smith, WL, 50, 111
Snyder, DL, 51
Space-charge limited electron stream, 48–9
Spontaneous emission, 79
Srinivasan, SK, 48, 50, 51, 97, 111, 143,
   158, 167, 171
Statistics:
   antibunched, 137, 142, 164, 166

Statistics: (*continued*)
   Bose–Einstein, 91, 93
   photoemission, 86, 99
   sub-Poissonian, 137, 164, 166
   super-Poissonian, 138
Stimulated emission, 79
Stoler, D, 142
Stratonovic, RL, 50
Subramanian, R, 51
Sudarshan, ECG, 2, 82, 84, 88
Sukavanam, S, 50, 158, 167
Superposition of fields, 162

Takahasi, H, 2, 110
Teich, MC, 142, 143
Titchmarsh, EC, 118

Tornau, CH, 2, 88, 110
Twiss, RQ, 2
Two-state process, 15

Vasudevan, R, 111, 143
Vere-Jones, D, 50, 51
Vernon, FL, 2
Von Mises, R, 49

Walker, JG, 142
Watson, HW, 69
Wolf, E, 2, 82, 88, 167

Yuan, JM, 142

Zeiger, HZ, 2, 88